SPACE SHUTTLE

Fact Archive

by Robert Godwin

CONTENTS

We acknowledge the financial support of the Government of Canada through the Book Publishing Industry Development Program for our publishing activities.

Published by Apogee Books, Box 62034, Burlington,
Ontario, Canada, L7R 4K2, http://www.apogeebooks.com
Tel: 905 637 5737
Printed and bound in Canada
Space Shuttle Fact Archive by Robert Godwin
ISBN 1-894959-52-3
ISBN-13 978-1894959-52-0

Shuttle Basics

The Space Shuttle is the world's first reusable spacecraft, and the first spacecraft in history that can carry large satellites both to and from orbit. The Shuttle launches like a rocket, maneuvers in Earth orbit like a spacecraft and lands like an airplane. Each of the three Space Shuttle orbiters now in operation — Discovery, Atlantis and Endeavour — is designed to fly at least 100 missions. So far, altogether they have flown a combined total of less than one-fourth of that.

Columbia was the first Space Shuttle orbiter to be delivered to NASA's Kennedy Space Center, Fla., in March 1979. Columbia and the STS-107 crew were lost Feb. 1, 2003, during re-entry. The Orbiter Challenger was delivered to KSC in July 1982 and was destroyed in an explosion during ascent in January 1986. Discovery was delivered in November 1983. Atlantis was delivered in April 1985. Endeavour was built as a replacement following the Challenger accident and was delivered to Florida in May 1991. An early Space Shuttle Orbiter, the Enterprise, never flew in space but was used for approach and landing tests at the Dryden Flight Research Center and several launch pad studies in the late 1970s. The Enterprise is now at the Smithsonian Air & Space Museum.

The Space Shuttle consists of three major components: the Orbiter which houses the crew; a large External Tank that holds fuel for the main engines; and two Solid Rocket Boosters which provide most of the Shuttle's lift during the first two minutes of flight. All of the components are reused except for the external fuel tank, which burns up in the atmosphere after each launch.

The longest the Shuttle has stayed in orbit on any single mission is 17.5 days on mission STS-80 in November 1996. Normally, missions may be planned for anywhere from five to 16 days in duration. The smallest crew ever to fly on the Shuttle numbered two people on the first few missions. The largest crew numbered eight people. Normally, crews may range in size from five to seven people. The Shuttle is designed to reach orbits ranging from about 185 kilometers to 643 kilometers (115 statute miles to 400 statute miles) high.

The Shuttle has the most reliable launch record of any rocket now in operation. Since 1981, it has boosted more than 1.36 million kilograms (3 million pounds) of cargo into orbit. More than than 600 crew members have flown on its missions. Although it has been in operation for almost 20 years, the Shuttle has continually evolved and is significantly different today than when it first was launched. NASA has made literally thousands of major and minor modifications to the original design that have made it safer, more reliable and more capable today than ever before.

Since 1992 alone, NASA has made engine and system improvements that are estimated to have tripled the safety of flying the Space Shuttle, and the number of problems experienced while a Space Shuttle is in flight has decreased by 70 percent. During the same period, the cost of operating the Shuttle has decreased by one and a quarter billion dollars annually — a reduction of more than 40 percent. At the same time, because of weight reductions and other improvements, the cargo the Shuttle can carry has increased by 7.3 metric tons (8 tons.)

In managing and operating the Space Shuttle, NASA holds the safety of the crew as its highest priority.

Solid Rocket Boosters

Solid Rocket Boosters separate The Solid Rocket Boosters (SRBs) operate in parallel with the main engines for the first two minutes of flight to provide the additional thrust needed for the Orbiter to escape the gravitational pull of the Earth. At an altitude of approximately 45 km (24 nautical miles), the boosters separate from the orbiter/external tank, descend on parachutes, and land in the Atlantic Ocean. They are recovered by ships, returned to land, and refurbished for reuse. The boosters also assist in guiding the entire vehicle during initial ascent. Thrust of both boosters is equal to 5,300,000 lbs.

In addition to the solid rocket motor, the booster contains the structural, thrust vector control, separation, recovery, and electrical and instrumentation subsystems.

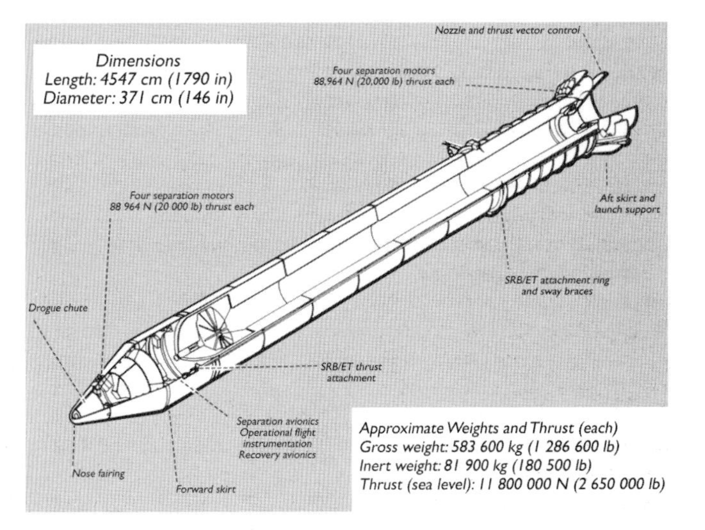

The solid rocket motor is the largest solid propellant motor ever developed for space flight and the first built to be used on a manned craft. The huge motor is composed of a segmented motor case loaded with solid propellants, an ignition system, a movable nozzle and the necessary instrumentation and integration hardware.

Each solid rocket motor contains more than 450,000 kg (1,000,000 lb.) of propellant, which requires an extensive mixing and casting operation at a plant in Utah. The propellant is mixed in 600 gallon bowls located in three different mixer buildings. The propellant is then taken to special casting buildings and poured into the casting segments.

Cured propellant looks and feels like a hard rubber typewriter eraser. The combined polymer and its curing agent is a synthetic rubber. Flexibility of the propellant is controlled by the ratio of binder to curing agent and the solid ingredients, namely oxidizer and aluminum. The solid fuel is actually powdered aluminum — a form similar to the foil wraps in your kitchen — mixed with oxygen provided by a chemical called ammonium perchlorate.

SRB Stats

Thrust at lift-off: 2,650,000 pounds
Propellant Properties:
    16% Atomized aluminum powder (fuel)
    69.8% Ammonium perchlorate (oxidizer)
    0.2% Iron oxide powder (catalyst)
    12% Polybutadiene acrylic acid acrylonite (binder)
    2% Epoxy curing agent
Weight
    Empty: 193,000 pounds
    Propellant: 1,107,000 pounds
    Gross: 1,300,000 pounds

The External Tank

The External Tank, or ET, is the "gas tank" for the Orbiter; it contains the propellants used by the Space Shuttle Main Engines.

The tank is also the "backbone" of the Shuttle during the launch, providing structural support for attachment with the solid rocket boosters and orbiter.

The tank is the only component of the Space Shuttle that is not reused. Approximately 8.5 minutes into the flight, with its propellant used, the tank is jettisoned.

At liftoff, the External Tank absorbs the total (7.8 million pounds) thrust loads of the three main engines and the two solid rocket motors.

When the Solid Rocket Boosters separate at an altitude of approximately 45 kilometers (28 miles), the orbiter, with the main engines still burning, carries the external tank piggyback to near orbital velocity, approximately 113 kilometers (70 miles) above the Earth. The now nearly empty tank separates and falls in a preplanned trajectory with the majority of it disintegrating in the atmosphere and the rest falling into the ocean.

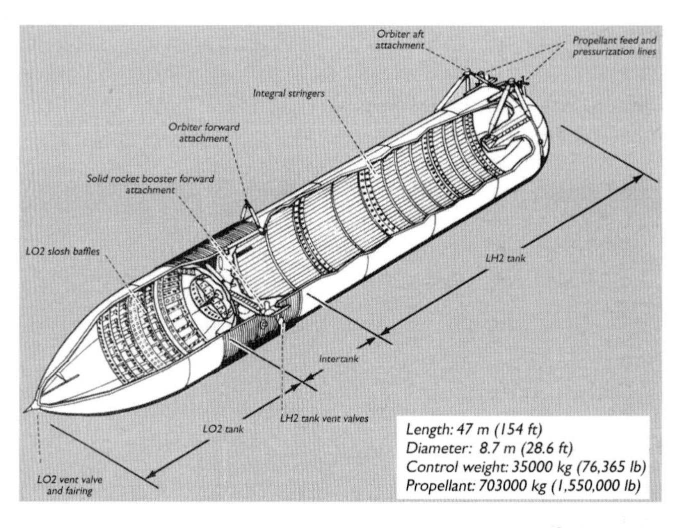

The three main components of the External Tank are an oxygen tank, located in the forward position, an aft-positioned hydrogen tank, and a collar-like intertank, which connects the two propellant tanks, houses instrumentation and processing equipment, and provides the attachment structure for the forward end of the solid rocket boosters.

The hydrogen tank is 2.5 times larger than the oxygen tank but weighs only one-third as much when filled to capacity. The reason for the difference in weight is that liquid oxygen is 16 times heavier than liquid hydrogen.

The skin of the External Tank is covered with a thermal protection system that is a 2.5-centimeter (1-inch) thick coating of spray-on polyisocyanurate foam. The purpose of the thermal protection system is to maintain the propellants at an acceptable temperature, to protect the skin surface from aerodynamic heat and to minimize ice formation.

The External Tank includes a propellant feed system to duct the propellants to the Orbiter engines, a pressurization and vent system to regulate the tank pressure, an environmental conditioning system to regulate the temperature and render the atmosphere in the intertank area inert, and an electrical system to distribute power and instrumentation signals and provide lightning protection.

The tank's propellants are fed to the Orbiter through a 43-centimeter (17-inch) diameter connection that branches inside the orbiter to feed each main engine.

External Tank Stats

Weight:
  Empty: 78,100 pounds
  Propellant: 1,585,379 pounds
  Gross: 1,667,677 pounds
Propellant Weight *
  Liquid oxygen: 1,359,142 pounds
  Liquid hydrogen: 226,237 pounds
  Gross: 1,585,379 pounds
Propellant Volume *
  Liquid oxygen tank: 143,060 gallons
  Liquid hydrogen tank: 383,066 gallons
  Gross: 526,126 gallons

* Liquid oxygen is 16 times heavier than liquid hydrogen.

Space Shuttle Main Engines

The three Space Shuttle Main Engines, in conjunction with the Solid Rocket Boosters, provide the thrust to lift the Orbiter off the ground for the initial ascent. The main engines continue to operate for 8.5 minutes after launch, the duration of the Shuttle's powered flight.

After the solid rockets are jettisoned, the main engines provide thrust which accelerates the Shuttle from 4,828 kilometers per hour (3,000 mph) to over 27,358 kilometers per hour (17,000 mph) in just six min-

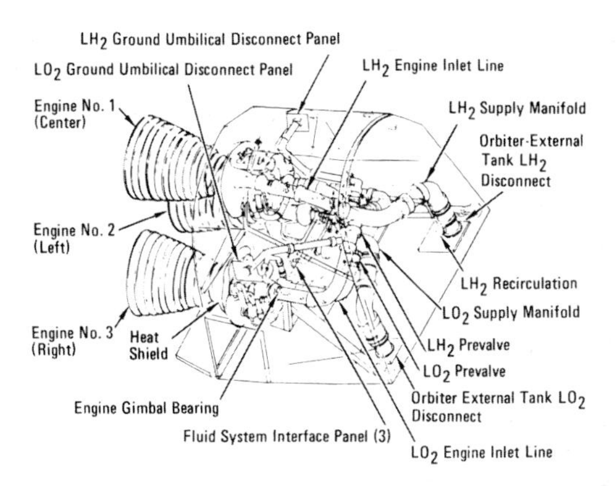

Main Propulsion System

utes to reach orbit. They create a combined maximum thrust of more than 1.2 million pounds.

As the Shuttle accelerates, the main engines burn a half-million gallons of liquid propellant provided by the large, orange external fuel tank. The main engines burn liquid hydrogen — the second coldest liquid on Earth at minus 423 degrees Fahrenheit (minus 252.8 degrees Celsius) — and liquid oxygen.

The engines' exhaust is primarily water vapor as the hydrogen and oxygen combine. As they push the Shuttle toward orbit, the engines consume liquid fuel at a rate that would drain an average family swimming pool in under 25 seconds generating over 37 million horsepower. Their turbines spin almost 13 times as fast as an automobile engine spins when it is running at highway speed.

The main engines develop thrust by using high-energy propellants in a staged combustion cycle. The propellants are partially combusted in dual preburners to produce high-pressure hot gas to drive the turbopumps.

Combustion is completed in the main combustion chamber. Temperatures in the main engine combustion chamber can reach as high as 6,000 degrees Fahrenheit (3,315.6 degrees Celsius).

Each Space Shuttle Main Engine operates at a liquid oxygen/liquid hydrogen mixture ratio of 6 to 1 to produce a sea level thrust of 179,097 kilograms (375,000 pounds) and a vacuum thrust of 213,188 (470,000 pounds).

The engines can be throttled over a thrust range of 65 percent to 109 percent, which provides for a high thrust level during liftoff and the initial ascent phase but allows thrust to be reduced to limit acceleration to 3 g's during the final ascent phase. The engines are gimbaled to provide pitch, yaw and roll control during the ascent.

Main Engine Stats

Thrust
    Sea level: 375,000 pounds
    Vacuum: 470,000 pounds
Nominal operating time: 8.5 minutes after liftoff
Propellant Mixture: 6 parts liquid oxygen to 1 part liquid hydrogen (by weight)
Weight: Approximately 6,700 pounds each
Dimensions: 14 feet long, 7.5 feet wide at mouth of nozzle
Life: 7.5 hours, 55 starts

The Orbiter

The Orbiter is both the brains and heart of the Space Transportation System. About the same size and weight as a DC-9 aircraft, the Orbiter contains the pressurized crew compartment (which can normally carry up to seven crew members), the huge cargo bay, and the three main engines mounted on its aft end.

The cockpit, living quarters and experiment operator's station are located in the forward fuselage of the Orbiter vehicle. Payloads are

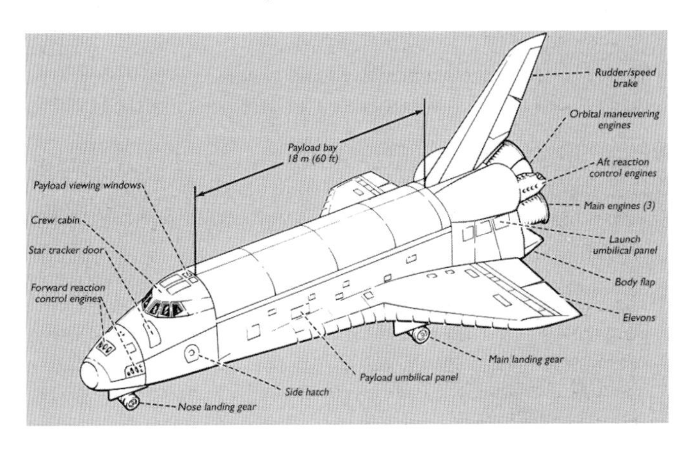

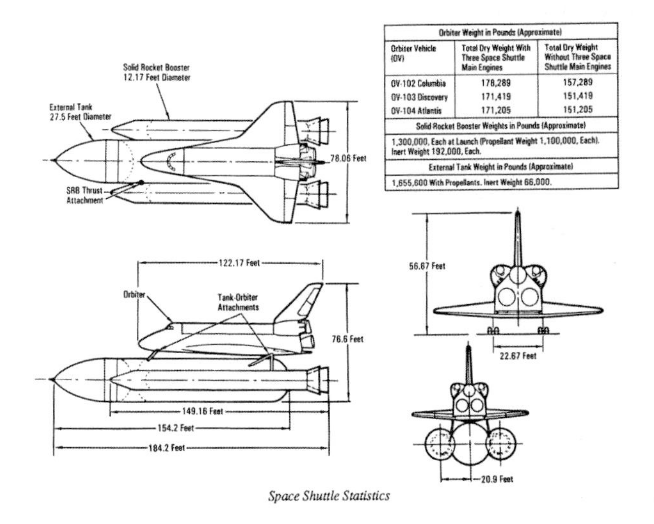

Orbiter Weight in Pounds (Approximate)		
Orbiter Vehicle (OV)	Total Dry Weight With Three Space Shuttle Main Engines	Total Dry Weight Without Three Space Shuttle Main Engines
OV-102 Columbia	178,289	157,289
OV-103 Discovery	171,419	151,419
OV-104 Atlantis	171,205	151,205
Solid Rocket Booster Weights in Pounds (Approximate)		
1,300,000, Each at Launch (Propellant Weight 1,100,000, Each). Inert Weight 192,000, Each.		
External Tank Weight in Pounds (Approximate)		
1,655,600 With Propellants. Inert Weight 66,000.		

Space Shuttle Statistics

carried in the mid-fuselage payload bay, and the Orbiter's main engines and maneuvering thrusters are located in the aft fuselage.

Forward Fuselage

The cockpit, living quarters and experiment operator's station are located in the forward fuselage. This area houses the pressurized crew module and provides support for the nose section, the nose gear and the nose gear wheel well and doors.

Crew Module

The 65.8-cubic-meter (2,325-cubic-foot) crew station module is a three-section pressurized working, living and stowage compartment in the forward portion of the Orbiter. It consists of the flight deck, the middeck/equipment bay and an airlock. Outside the aft bulkhead of the crew module in the payload bay, a docking module and a transfer tunnel with an adapter can be fitted to allow crew and equipment transfer for docking, Spacelab and extravehicular operations.

The two-level crew module has a forward flight deck with the commander's seat positioned on the left and the pilot's seat on the right.

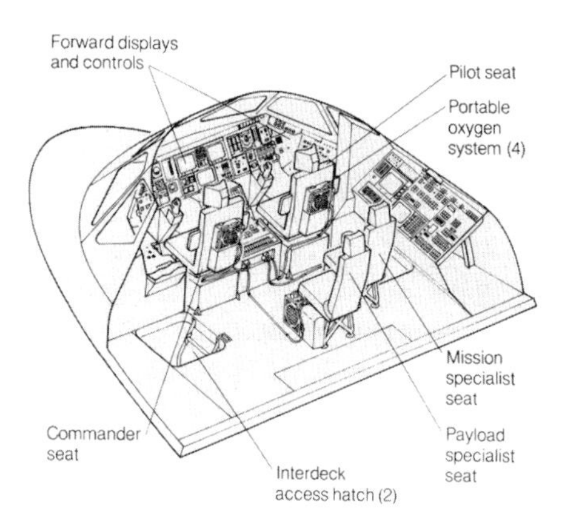

Flight deck looking forward

Flight Deck

The flight deck is designed in the usual pilot/copilot arrangement, which permits the vehicle to be piloted from either seat and permits one-man emergency return. Each seat has manual flight controls, including rotation and translation hand controllers, rudder pedals and speed-brake controllers. The flight deck seats four. The on-orbit displays and controls are at the aft end of the flight deck/crew compartment. The displays and controls on the left are for operating the Orbiter, and those on the right are for operating and handling the payloads. More than 2,020 separate displays and controls are located on the flight deck.

Six pressure windshields, two overhead windows and two rear-viewing payload bay windows are located in the upper flight deck of the crew module, and a window is located in the crew entrance/exit hatch located in the midsection, or deck, of the crew module.

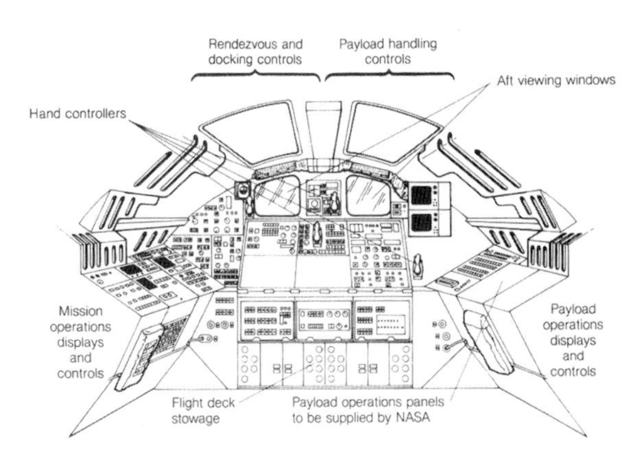

FLIGHT DECK
looking aft

Middeck

The middeck contains provisions and stowage facilities for four crew sleep stations. Stowage for the lithium hydroxide canisters and other gear, the waste management system, the personal hygiene station and the work/dining table is also provided in the middeck.

The nominal maximum crew size is seven. The middeck can be reconfigured by adding three rescue seats in place of the modular stowage and sleeping provisions. The seating capacity will then accommodate the rescue flight crew of three and a maximum rescued crew of seven.

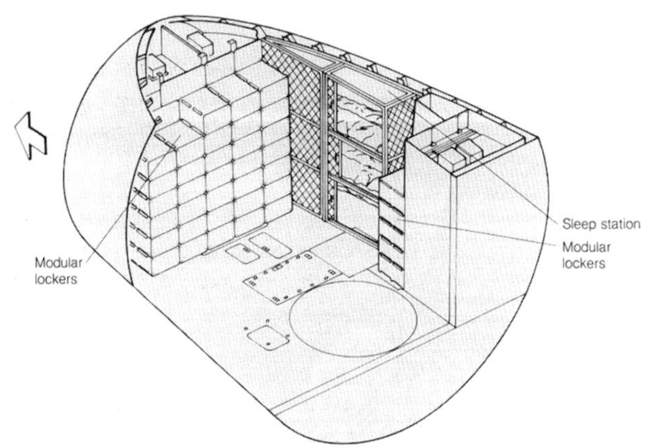

Mid deck looking forward

Airlock

The airlock provides access for spacewalks, known as extravehicular activity, or EVA. It can be located in one of several places: inside the Orbiter crew module in the middeck area mounted to the aft bulkhead, outside the cabin also mounted to the bulkhead or on top of a tunnel adapter that can connect the pressurized Spacehab module with the Orbiter cabin. A docking module can also serve as an EVA airlock.
The airlock contains two spacesuits, expendables for two six-hour payload EVAs and one contingency or emergency EVA, and mobility aids such as handrails to enable the crew to perform a variety of tasks.

The airlock allows two crewmen room for changing spacesuits.

Midfuselage

In addition to forming the payload bay of the Orbiter, the midfuselage supports the payload bay doors, hinges and tiedown fittings, the forward wing glove and various Orbiter system components.

Each payload bay door supports four radiator panels. When the doors are opened, the tilting radiators are unlatched and moved to the proper position. This allows heat radiation from both sides of the panels, whereas the four aft radiator panels radiate from the upper side only.

Some payloads may not be attached directly to the Orbiter but to payload carriers that are attached to the Orbiter. The inertial upper stage, pressurized modules or any specialized cradle for holding a payload are typical carriers.

The Remote Manipulator System, or RMS, is a 15.2-meter (50-foot) long articulating arm remotely controlled from the flight deck of the Orbiter. The elbow and wrist movements permit payloads to be grappled for deployment out of the payload bay or retrieved and secured for return to Earth.

A television camera and lights near the outer end of the arm permit the operator to see on television monitors what his hands are doing. In addition, three floodlights are located along each side of the payload bay.

Aft Fuselage

The aft fuselage consists of the left and right orbital maneuvering systems, Space Shuttle main engines, body flap, vertical tail and Orbiter/external tank rear attachments.

The forward bulkhead closes off the aft fuselage from the midfuselage. The upper portion of the bulkhead attaches to the vertical tail. The internal thrust structure supports the three Space Shuttle main engines, low pressure turbopumps and propellant lines.

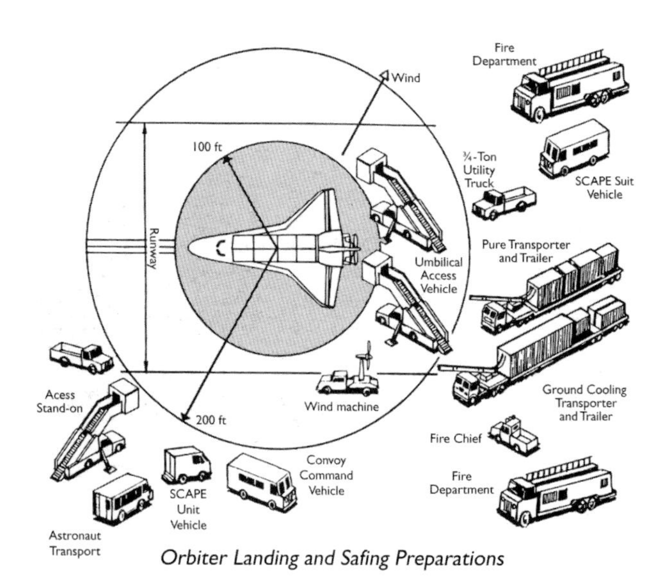

Orbiter Landing and Safing Preparations

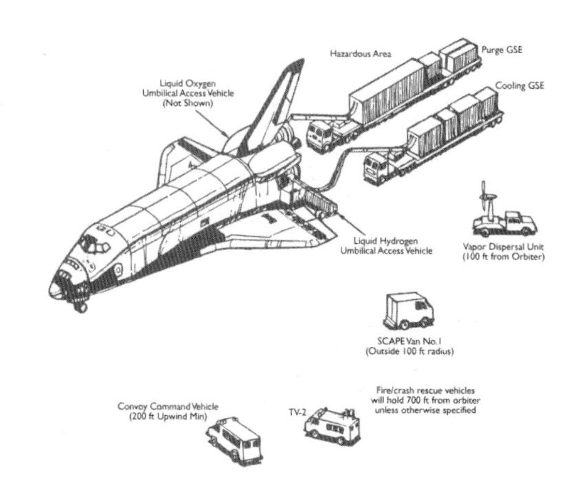

Orbiter Safing Operations

Orbiter Stats:

Height: (on runway) 57 feet
Length: 122 feet
Wingspan: 78 feet
Mid Fuselage:
    Length: 60 feet
    Width: 17 feet
    Height: 13 feet
Aft Fuselage
    Length: 18 feet
    Width: 22 feet
    Height:20 feet
Payload Bay Doors
    Length: 60 feet
    Diameter: 15 feet
    Width: 22.67 feet
    Surface: 1,600 feet²

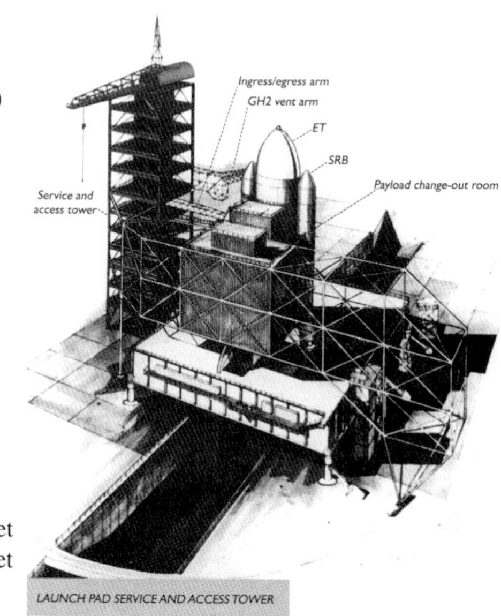

LAUNCH PAD SERVICE AND ACCESS TOWER

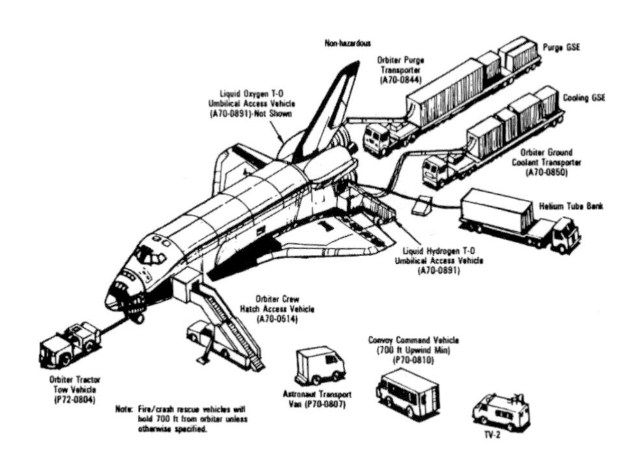

Orbiter Power-Down and Crew Egress

NASA ORBITER FLEET

Space Shuttle Overview: Atlantis (OV-104)

NASA's fourth space-rated Space Shuttle orbiter, OV-104 "Atlantis," was named after the two-masted boat that served as the primary research vessel for the Woods Hole Oceanographic Institute in Massachusetts from 1930 to 1966. The boat had a 17-member crew and accommodated up to five scientists who worked in two onboard laboratories, examining water samples and marine life. The crew also used the first electronic sounding devices to map the ocean floor.

Construction of the orbiter Atlantis began on March 3, 1980. Thanks to lessons learned in the construction and testing of orbiters Enterprise, Columbia and Challenger, Atlantis was completed in about half the time in man-hours spent on Columbia. This is largely attributed to the use of large thermal protection blankets on the orbiter's upper body, rather than individual tiles requiring more attention.

Weighing in at 151,315 pounds when it rolled out of the assembly plant in Palmdale, Calif., Atlantis was nearly 3.5 tons lighter than Columbia. The new orbiter arrived at NASA's Kennedy Space Center in Florida on April 9, 1985, and over the next seven months was prepared for her maiden voyage.

Like her seafaring predecessor, orbiter Atlantis has carried on the spirit of exploration with several important missions of her own. On Oct. 3, 1985, Atlantis launched on her first space flight, STS-51-J, with a classified payload for the U.S. Department of Defense. The vehicle went on to carry four more DOD payloads on later missions.

Atlantis also served as the on-orbit launch site for many noteworthy spacecraft, including planetary probes Magellan and Galileo, as well as the Compton Gamma Ray Observatory. An impressive array of onboard science experiments took place during most missions to further enhance space research in low Earth orbit.

Orbiter OV-104 *Atlantis*
lands at Edwards Air Force Base

Starting with STS-71, Atlantis pioneered the Shuttle-Mir missions, flying the first seven missions to dock with the Russian space station. When linked, Atlantis and Mir together formed the largest spacecraft in orbit at the time. The missions to Mir included the first on-orbit U.S. crew exchanges, now a common occurrence on the International Space Station. On STS-79, the fourth docking mission, Atlantis ferried astronaut Shannon Lucid back to Earth after her record-setting 188 days in orbit aboard Mir.

In recent years, Atlantis has delivered several vital components to the International Space Station, including the U.S. laboratory module, Destiny, as well as the Joint Airlock Quest and multiple sections of the Integrated Truss structure that makes up the Station's backbone. As NASA seeks to fulfill the Vision for Space Exploration, beginning with the completion of the Station, Atlantis will be called upon for many missions to come.

Construction Milestones - OV-104

Jan. 29, 1979	Contract Award
March 30, 1980	Start structural assembly of crew module
Nov. 23, 1981	Start structural assembly of aft-fuselage
June 13, 1983	Wings arrive at Palmdale from Grumman
Dec. 2, 1983	Start of Final Assembly
April 10, 1984	Completed final assembly
March 6, 1985	Rollout from Palmdale
April 3, 1985	Overland transport from Palmdale to Edwards
April 9, 1985	Delivery to Kennedy Space Center
Sept. 5, 1985	Flight Readiness Firing
Oct. 3, 1985	First Flight (STS-51-J)

Upgrades and Features

By early 2005, Atlantis had undergone two overhauls known as Orbiter Maintenance Down Periods. Some of the most significant upgrades and new features included:

* Installation of the drag chute
* New plumbing lines and electrical connections configuring the orbiter for extended duration missions
* New insulation for the main landing gear doors
* Improved nosewheel steering
* Preparations for the Mir Orbiter Docking System unit later installed at Kennedy
* Installation of the International Space Station airlock and Orbiter Docking System
* Installation of the Multifunction Electronic Display System, or "glass cockpit"

Space Shuttle Overview: Challenger (OV-099)

First called STA-099, Challenger was built to serve as a test vehicle for the Space Shuttle program. But despite its Earth-bound beginnings, STA-099 was destined for space.

Orbiter OV-099 *Challenger* in orbit with payload bay doors open and Canadarm deployed

In the late 1970s, NASA strived for a lighter weight orbiter, but a test vehicle was needed to ensure the lighter airframe could handle the stress of space flight. Computer software at the time wasn't yet advanced enough to accurately predict how STA-099's new, optimized design would respond to intense heat and stress. The best solution was to submit the vehicle to a year of intensive vibration and thermal testing.

In early 1979, NASA awarded Space Shuttle orbiter manufacturer Rockwell a contract to convert STA-099 to a space-rated orbiter, OV-099. The vehicle's conversion began late that year. Although the job was easier than it would have been to convert NASA's first orbiter, Enterprise, it was a major process that involved the disassembly and replacement of many parts and components.

The second orbiter to join NASA's Space Shuttle fleet, OV-099 arrived at NASA's Kennedy Space Center in Florida in July 1982, bearing the name "Challenger."

Space Shuttle orbiter Challenger was named after the British Naval research vessel HMS Challenger that sailed the Atlantic and Pacific oceans during the 1870s. The Apollo 17 lunar module also carried the name of Challenger. Like its historic predecessors, Challenger and her crews made significant scientific contributions in the spirit of exploration.

Challenger launched on her maiden voyage, STS-6, on April 4, 1983. That mission saw the first spacewalk of the Space Shuttle program, as well as the deployment of the first satellite in the Tracking and Data Relay System constellation. The orbiter launched the first American woman, Sally Ride, into space on mission STS-7 and was the first to carry two U.S. female astronauts on mission STS 41-G.

The first orbiter to launch and land at night on mission STS-8, Challenger also made the first Space Shuttle landing at Kennedy Space Center, concluding mission STS 41-B. Spacelabs 2 and 3 flew aboard the ship on missions STS 51-F and STS 51-B, as did the first German-dedicated Spacelab on STS 61-A. A host of scientific experiments and satellite deployments were performed during Challenger's missions.

Challenger's service to America's Space Program ended in tragedy on Jan. 28, 1986. Just 73 seconds into mission STS 51-L, a booster failure caused an explosion that resulted in the loss of seven astronauts, as well as the vehicle.

The loss of Challenger does not overshadow her legacy in NASA's storied history. The discoveries made on her many successful missions continue to better mankind in space flight and in life on Earth.

Construction Milestones - STA-099

July 26, 1972	Contract Award
Nov. 21, 1975	Start structural assembly of crew module
June 14, 1976	Start structural assembly of aft-fuselage
March 16, 1977	Wings arrive at Palmdale from Grumman
Sept. 30, 1977	Start of Final Assembly
Feb. 10, 1978	Completed final assembly
Feb. 14, 1978	Rollout from Palmdale

Construction Milestones - OV-099

Jan. 1, 1979	Contract Award
Jan. 28, 1979	Start structural assembly of crew module
June 14, 1976	Start structural assembly of aft-fuselage
March 16, 1977	Wings arrive at Palmdale from Grumman
Nov. 3, 1980	Start of Final Assembly
Oct. 21, 1981	Completed final assembly
June 30, 1982	Rollout from Palmdale
July 1, 1982	Overland transport from Palmdale to Edwards
July 5, 1982	Delivery to Kennedy Space Center
Dec. 19, 1982	Flight Readiness Firing
April 4, 1983	First Flight (STS-6)

Space Shuttle Overview: Columbia (OV-102)

Columbia lifts off from the launch pad for the very first time. On April 12, 1981, a bright white Columbia roared into a deep blue sky as the nation's first reusable Space Shuttle. Named after the first American ocean vessel to circle the globe and the command module

Orbiter OV-102 *Columbia*
lands at the Kennedy Space Centre

for the Apollo 11 Moon landing, Columbia continued this heritage of intrepid exploration. The heaviest of NASA's orbiters, Columbia weighed too much and lacked the necessary equipment to assist with assembly of the International Space Station. Despite its limitations, the orbiter's legacy is one of groundbreaking scientific research and notable "firsts" in space flight.

Space Shuttle mission STS-9, launched in late November 1983, was the maiden flight for Spacelab. Designed to be a space-based science lab, Spacelab was installed inside the orbiter's cargo bay. Spacelab featured an enclosed crew work module connected to an outside payload pallet, which could be mounted with various instruments and experiments. From inside the lab, astronauts worked with the experiments on the pallet and within the crew module itself. The lab would go on to fly aboard the rest of the fleet, playing host throughout its accomplished lifetime to unprecedented research in astronomy, biology and other sciences. Spacelab ultimately finished where its career began; its 16th and final mission was hoisted into space aboard Columbia in 1998.

In addition to Columbia's STS-1 flight and Spacelab, the orbiter was also the stage for many other remarkable firsts. Germany's Dr. Ulf Merbold became the first European Space Agency astronaut when he flew aboard 1983's STS-9. The Japanese Space Agency and STS-65's Chiaki Mukai entered history as the first Japanese woman to fly in space in 1994. In a display of national pride, the crew of STS-73 even "threw" the ceremonial first pitch for game five of the 1995 baseball World Series, marking the first time the pitcher was not only outside of the stadium, but out of this world.

Perhaps Columbia's crowning achievement was the deployment of the gleaming Chandra X-ray Observatory in July 1999. Carried into space inside the orbiter's payload bay, the slender and elegant Chandra telescope was released on July 23. Still in flight today, the X-ray telescope specializes in viewing deep space objects and finding the answers to astronomy's most fundamental questions.

Columbia and its crew were tragically lost during STS-107 in 2003.

As the Space Shuttle lifted off from Kennedy Space Center in Florida on January 16, a small portion of foam broke away from the orange external fuel tank and struck the orbiter's left wing. The resulting damage created a hole in the wing's leading edge, which caused the vehicle to break apart during reentry to Earth's atmosphere on February 1.

Construction Milestones

July 26, 1972	Contract Award
March 25, 1975	Start long lead fabrication aft fuselage
November 17, 1975	Start long-lead fabrication of crew module
June 28, 1976	Start assembly of crew module
September 13, 1976	Start structural assembly of aft-fuselage
December 13, 1976	Start assembly upper forward fuselage
January 3, 1977	Start assembly vertical stabilizer
August 26, 1977	Wings arrive at Palmdale from Grumman
October 28, 1977	Lower forward fuselage on dock, Palmdale
November 7, 1977	Start of Final Assembly
February 24, 1978	Body flap on dock, Palmdale
April 28, 1978	Forward payload bay doors on dock, Palmdale
May 26,1978	Upper forward fuselage mate
July 7, 1978	Complete mate forward and aft payload bay doors
September 11, 1978	Complete forward RCS
February 3, 1979	Complete combined systems test, Palmdale
February 16, 1979	Airlock on dock, Palmdale
March 5, 1979	Complete postcheckout
March 8, 1979	Closeout inspection, Final Acceptance Palmdale
March 8, 1979	Rollout from Palmdale to Dryden (38 miles)
March 12, 1979	Overland transport from Palmdale to Edwards
March 20, 1979	SCA Ferry Flight from DFRF to Bigs AFB, Tx
March 22, 1979	SCA Ferry flight from Bigs AFB to Kelly AFB, Tx
March 24, 1979	SCA Ferry flight from Kelly AFB to Eglin AFB, FL
March 24, 1979	SCA Ferry flight from Eglin, AFB to KSC
November 3, 1979	Auxiliary Power Unit hot fire tests, OPF KSC
December 16, 1979	Orbiter integrated test start, KSC
January 14, 1980	Orbiter integrated test complete, KSC
February 20, 1981	Flight Readiness Firing
April 12, 1981	First Flight (STS-1)

Upgrades and Features

Columbia is commonly referred to as OV-102, for Orbiter Vehicle-102. The orbiter weighed 178,000 pounds with its main engines installed.
Columbia was the first orbiter to undergo the scheduled inspection and retrofit program. In 1991, Columbia returned to its birthplace at Rockwell International's Palmdale, Calif., assembly plant. The spacecraft underwent approximately 50 upgrades there, including the addition of carbon brakes and a drag chute, improved nose wheel steering, removal of instrumentation used during the test phase of the orbiter, and an enhancement of its Thermal Protection System. The orbiter returned to Florida in February 1992 to begin processing for mission STS-50, launching in June of that year.

In 1994, Columbia was transported back to Palmdale for its first major tear-down and overhaul, known as the Orbiter Maintenance Down Period (OMDP). This overhaul typically lasts one year or longer and leaves the vehicle in "like-new" condition.

Its second OMDP came in 1999, when workers performed more than 100 modifications on the vehicle. The orbiter's most impressive upgrade likely was the installation of a state-of-the-art, Multi-functional Electronic Display System (MEDS), or "glass cockpit." The MEDS replaced traditional instrument dials and gauges with small, computerized video screens. The new system improved crew interaction with the orbiter during flight and reduced maintenance costs by eliminating the outdated and tricky electromechanical displays.

Space Shuttle Overview: Discovery (OV-103)

Discovery (OV-103) was NASA's third space shuttle orbiter to join the fleet, arriving for the first time at the Kennedy Space Center in Florida in November 1983.

After checkout and processing, it was launched on Aug. 30, 1984, for its first mission, 41-D, to deploy three communications satellites.

Orbiter OV-103 *Discovery*
in transit from California to Florida
on the NASA transporter plane

Since that inaugural flight, Discovery has completed more than 30 successful missions, surpassing the number of flights made by any other orbiter in NASA's fleet. Just like all of the orbiters, it has undergone some major modifications over the years. The most recent began in 2002 and was the first carried out at Kennedy. It provided 99 upgrades and 88 special tests, including new changes to make it safer for flight.

Discovery has the distinction of being chosen as the Return to Flight orbiter twice. The first was for STS-26 in 1988, and the second when it carried the STS-114 crew on NASA's Return to Flight mission to the International Space Station in July 2005.

The choice of the name "Discovery" carried on a tradition drawn from some historic, Earth-bound exploring ships of the past. One of these sailing forerunners was the vessel used in the early 1600s by Henry Hudson to explore Hudson Bay and search for a northwest passage from the Atlantic to the Pacific.

Another such ship was used by British explorer James Cook in the 1770s during his voyages in the South Pacific, leading to the discovery of the Hawaiian Islands. In addition, two British Royal Geographical Society ships have carried the name "Discovery" as they sailed on expeditions to the North Pole and the Antarctic.

Destined for exploring the heavens instead of the seas, it was only fitting that NASA's Discovery carried the Hubble Space Telescope into space during mission STS-31 in April 1990, and provided both the second and third Hubble servicing missions (STS-82 in February 1997 and STS-103 in December 1999).

During its many successful trips to space, Discovery has carried satellites aloft, ferried modules and crew to the International Space Station, and provided the setting for countless scientific experiments.

Construction Milestones

January 29, 1979	Contract Award
August 27, 1979	Start long lead fabrication of Crew Module
June 20, 1980	Start fabrication lower fuselage
November 10, 1980	Start structural assembly of aft-fuselage
December 8, 1980	Start initial system installation aft fuselage
March 2, 1981	Start fabrication/assembly of payload bay doors
October 26, 1981	Start initial system installation, crew module, Downey
January 4, 1982	Start initial system installation upper forward fuselage
March 16, 1982	Midfuselage on dock, Palmdale
March 30, 1982	Elevons on dock, Palmdale
April 30, 1982	Wings arrive at Palmdale from Grumman
April 30, 1982	Lower forward fuselage on dock, Palmdale
July 16, 1982	Upper forward fuselage on dock, Palmdale
August 5, 1982	Vertical stabilizer on dock, Palmdale
September 3, 1982	Start of Final Assembly
October 15, 1982	Body flap on dock, Palmdale
January 11, 1983	Aft fuselage on dock, Palmdale
February 25, 1983	Final assembly and closeout installation, Palmdale
February 28, 1983	Start initial subsystems test, power-on, Palmdale
May 13, 1983	Complete initial subsystems testing
July 26, 1983	Complete subsystems testing
August 12, 1983	Completed Final Acceptance

October 16, 1983	Rollout from Palmdale
November 5, 1983	Overland transport from Palmdale to Edwards
November 9, 1983	Delivery to Kennedy Space Center
June 2, 1984	Flight Readiness Firing
August 30, 1984	First Flight (41-D)

Upgrades and Features

Discovery benefited from lessons learned in the construction and testing of Enterprise, Columbia and Challenger. At rollout, its weight was some 6,870 pounds less than Columbia.

Beginning in the fall of 1995, the orbiter underwent a nine-month Orbiter Maintenance Down Period (OMDP) in Palmdale California. The vehicle was outfitted with a 5th set of cryogenic tanks and an external airlock to support missions to the International Space Station. It returned to the Kennedy Space Center, riding piggy-back on a modified Boeing 747, in June 1996.

Following STS-105, Discovery became the first of the orbiter fleet to undergo Orbiter Major Modification (OMM) period at the Kennedy Space Center. Work began in September 2002, and along with the scheduled upgrades, additional safety modifications were added as part of the preparations for Return to Flight.

Space Shuttle Overview: Endeavour (OV-105)

Authorized by Congress in August 1987 as a replacement for the Space Shuttle orbiter Challenger, Endeavour (OV-105) arrived at Kennedy Space Center's Shuttle Landing Facility on May 7, 1991, piggy-backed on top of NASA's new Space Shuttle Carrier Aircraft.

Space Shuttle Endeavour launch at Kennedy Space Center For the first time, an orbiter was named through a national competition involving students in elementary and secondary schools. They were asked to select a name based upon an exploratory or research sea vessel. In May 1989, President George Bush announced the winning name.

Orbiter OV-105 *Endeavour*
leaving Edwards Air Force Base for Florida

Endeavour was named after a ship chartered to traverse the South Pacific in 1768 and captained by 18th century British explorer James Cook, an experienced seaman, navigator and amateur astronomer. He commanded a crew of 93 men, including 11 scientists and artists.

Cook's main objective, tasked by the British Admiralty and the Royal Society, was to observe the Transit of Venus at Tahiti. This reading enabled astronomers to find the distance of the Sun from the Earth, which then could be used as a unit of measurement in calculating the parameters of the universe.

Cook's achievements on Endeavour were numerous, including the accurate charting of New Zealand and Australia and successfully navigating the Great Barrier Reef. Thousands of new plant specimens and animal species were observed and illustrated on this maiden voyage. Cook also established the usefulness of including scientists on voyages of exploration.

Space Shuttle Endeavour embodies similar experiences. Its first launch, the STS-49 mission, began with a flawless liftoff on May 7,

1992, beginning a journey filled with excitement, anticipation and many firsts.

One of Endeavour's primary assignments was to capture INTELSAT VI, an orbiting, but not functioning, communications satellite, and replace its rocket motor. Unfortunately, the Space Shuttle wasn't designed to retrieve the satellite, which created many repair challenges.

The project sparked public interest in the mission and NASA received a deluge of suggestions on possible ways for the crew to grab onto the satellite. It took three attempts to capture the satellite for repairs to be made. An unprecedented three-person spacewalk took place after the procedure was evaluated by the astronauts and ground team.

Between rescue attempts, the STS-49 crew was busy with a variety of activities. They conducted medical tests assessing the human body's performance in microgravity, and recorded footage for an educational video comparing Cook's first voyage on Endeavour with the Space Shuttle orbiter's maiden voyage.

Once the new motor was attached, it propelled the satellite into the correct orbit, providing a relay link for the equivalent of 120,000 two-way simultaneous telephone calls and three television channels.

This was the first time four spacewalks were conducted on a Space Shuttle mission and one of them was the longest in space history, lasting more than eight hours.

The crew also took part in the Commercial Protein Crystal Growth experiment. The research tested the production of protein crystals grown in microgravity.

Because of Endeavour's excellent performance, NASA decided to extend the flight two days to complete more mission objectives and allow the crew enough time to prepare for landing.

OV-105 became the first Space Shuttle orbiter to use a drag chute during a landing — only one of many technical improvements made to Endeavour.

Just as James Cook set the standard with his seafaring Endeavour voyage, the Space Shuttle Endeavour missions have continued to uphold and surpass the standards set by its namesake, more than 200 years later.

Construction Milestones

February 15, 1982	Start structural assembly of Crew Module
July 31, 1987	Contract Award
September 28, 1987	Start structural assembly of aft-fuselage
December 22, 1987	Wings arrive at Palmdale, Calif. from Grumman
August 1, 1987	Start of Final Assembly
July 6, 1990	Completed final assembly
April 25, 1991	Rollout from Palmdale
May 7, 1991	Delivery to Kennedy Space Center
April 6, 1992	Flight Readiness Firing
May 7, 1992	First Flight (STS-49)

Upgrades and Features

Spare parts from the construction of Discovery (OV-103) and Atlantis (OV-104), manufactured to facilitate the repair of an orbiter if needed, were eventually used to build OV-105.

Endeavour also featured new hardware, designed to improve and expand orbiter capabilities. Most of this equipment was later incorporated into the other three orbiters during out-of-service major inspection and modification programs.

Endeavour's upgrades include:

* A 40-foot-diameter drag chute that reduces the orbiter's rollout distance by 1,000 to 2,000 feet.
* An updated avionics system that include advanced general purpose computers, improved inertial measurement units and tactical air navigation systems, enhanced master events controllers and

multiplexer-demultiplexers, a solid-state star tracker.
* Improved nose wheel steering mechanisms.
* An improved version of the Auxiliary Power Units that provide power to operate the Space Shuttle's hydraulic systems.
* Installation of an external airlock, making Endeavour capable of docking with the International Space Station.
* Originally equipped as the first extended duration orbiter, later removed during OMDP to save weight for ISS missions.
* Installation of a ground cooling hookup to allow payload bay to cool the mini-pressurized logistics module (MPLM).
* General weight-reduction program to maximize the payload capability to the ISS.
* Doublers added to several wing spars to allow heavier payloads and two wing glove truss tubes were replaced with units having increased wall thickness.
* Approximately 100 modifications made to Endeavour during its first Orbiter Major Modification period (OMDP).

Space Shuttle Endeavour underwent OMDP, which began in December 2003. Engineers and technicians spent thousands of hours performing 124 modifications to the vehicle. These included recommended return to flight safety modifications, bonding more than 1,000 thermal protection system tiles and inspecting more than 150 miles of wiring.

Two of the more extensive modifications included the addition of the multi-functional electronic display system (glass cockpit), and the three-string global positioning system.

The glass cockpit is a new, full-color, flat-panel display system that improves interaction between the crew and orbiter. It provides easy-to-read graphics portraying key flight indicators like attitude display and mach speed. Endeavour was the last vehicle in the fleet to receive this system.

The three-string global positioning system will improve the shuttle's landing capability. It will allow Endeavour to make a landing at any runway long enough to handle the shuttle. The previous system only allowed for landings at military bases. Shuttle major modification periods are scheduled at regular intervals to enhance safety and performance, infuse new technology and allow thorough inspections of the airframe and wiring. This was the second of modification period performed entirely at Kennedy. Endeavour's previous modification was completed in March 1997.

SPACEHAB - Research Laboratory

The SPACEHAB Space Research Laboratory is located in the forward end of the Shuttle orbiter cargo bay and is accessed from the orbiter middeck through a tunnel adapter connected to the airlock. SPACEHAB weighs 9,628 pounds, is 9.2 feet long, 11.2 feet high and 13.5 feet in diameter. It increases pressurized experiment space in the Shuttle orbiter by 1100 cubic feet, quadrupling the working and storage volume available. Environmental control of the laboratory's interior maintains ambient temperatures between 65 and 80 degrees Fahrenheit.

The laboratory has a total payload capacity of 3000 pounds and in addition to facilitating crew access, provides experiments with services such as power, temperature control and command/data functions.

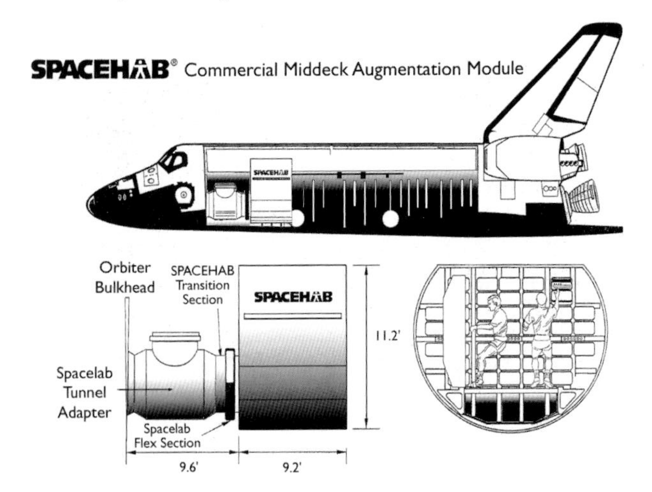

LOGISTICS SINGLE MODULE (SPACEHAB LSM)

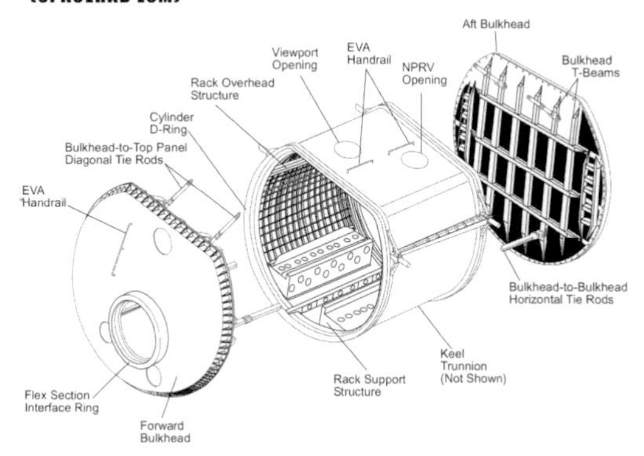

Other services, such as late access/early retrieval, also are available. The SPACEHAB Space Research Laboratory can provide various physical accommodations to users based on size, weight and other requirements. Experiments are commonly integrated into the laboratory in Shuttle middeck- type lockers or SPACEHAB racks. The laboratory can accommodate up to 61 lockers, with each locker providing a maximum capacity of 60 pounds and 2.0 cubic feet of volume.

SPACEHAB RESEARCH DOUBLE MODULE

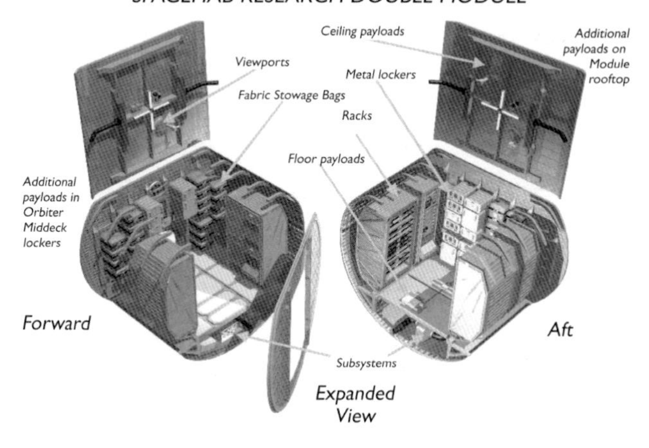

The laboratory also can accommodate up to two SPACEHAB racks, either of which can be a "double-rack" or "single-rack" configuration, but each rack used reduces the number of usable locker locations by 10 lockers. A "double- rack" provides a maximum capacity of 1250 pounds and 45 cubic feet of volume, whereas a "single-rack" provides half of that capacity. The "double-rack" is similar in size and design to the racks planned for use in the space station. The use of lockers or racks is not essential for integration into the SPACEHAB Space Research Laboratory. Payloads also can be accommodated by directly mounting them on the laboratory.

SPACEHAB's Logistics Double Module is a 20-foot long, 14-foot wide, 11.2-foot high pressurized aluminum module carried in the shuttle payload bay and connected to the middeck area of the orbiter by an access tunnel. Designed to augment the Shuttle's middeck, the double module has a total cargo capacity of up to 10,000 pounds and contains the systems necessary to support the crewmembers, such as ventilation, lighting and limited power.

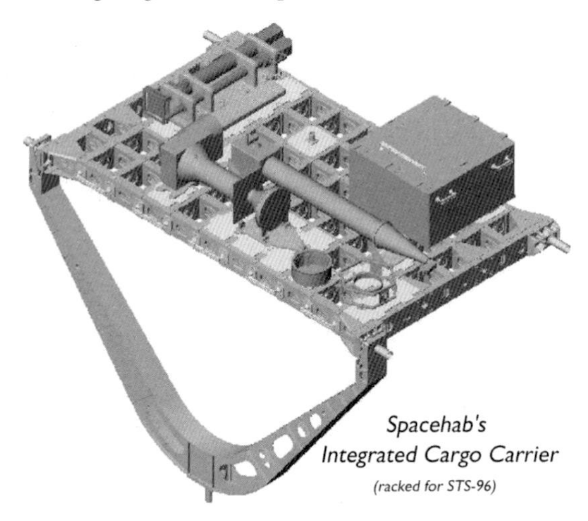
Spacehab's Integrated Cargo Carrier
(racked for STS-96)

SPACEHAB's Integrated Cargo Carrier is 8 feet long, 15 feet wide and 10 inches thick. It has a capacity of up to 6,000 pounds of attached payload. Cargo can be attached to both the top and bottom of the pallet. The Integrated Cargo Carrier is a two-part structure consisting of a waffle-like aluminum box-beam pallet and keel-yoke assembly. It is installed in the cargo bay of the space shuttle orbiter, to expand the shuttle's capacity to transport unpressurized cargo.

The SPACEHAB RDM was approximately 20 feet long, 14 feet wide, and 11 feet high. Outfitted as a state-of-the-art laboratory, it had a pressurized volume of 2,200 cubic feet and could hold up to 61 space shuttle middeck lockers (up to 60 pounds and 2 cubic feet each) plus six Double Racks (1,400 pounds and 45 cubic feet each). The RDM also could accommodate International Space Station Payload Racks (ISPRs). The Module had two viewports and could carry powered rooftop payloads using feed-through plates in the module ceiling. The RDM, which had a payload capacity of 9,000 pounds, was carrying about 7,500 pounds of research payloads on STS-107.

SPACELAB - Research Laboratory

Spacelab is a reusable, research laboratory facility. When carried in the payload bay of the Space Shuttle orbiter, it converts the Shuttle into a versatile, on orbit research center. Modular in design and construction, Spacelab consists of several interchangeable components that can be assembled in different ways to meet the needs of a particular scientific research mission. Spacelab consists of two major ele-

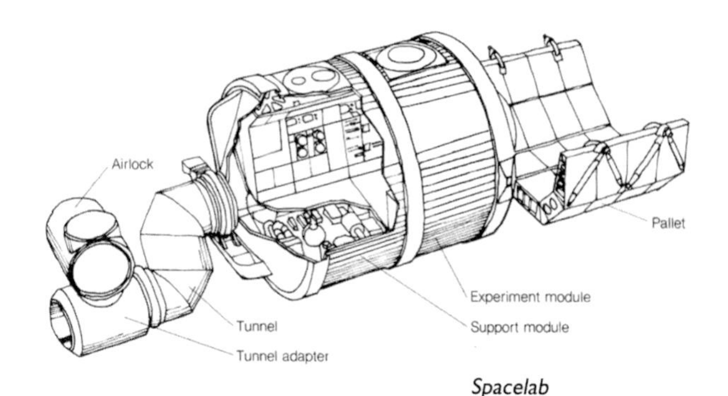

Spacelab

SPACELAB 1 MODULE

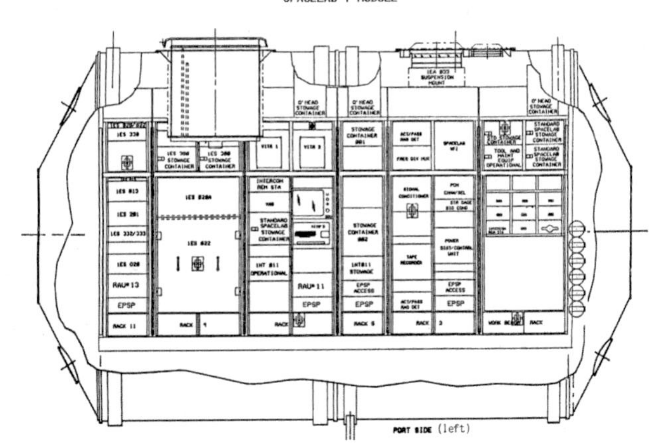

FORWARD →

SPACELAB 1 MODULE

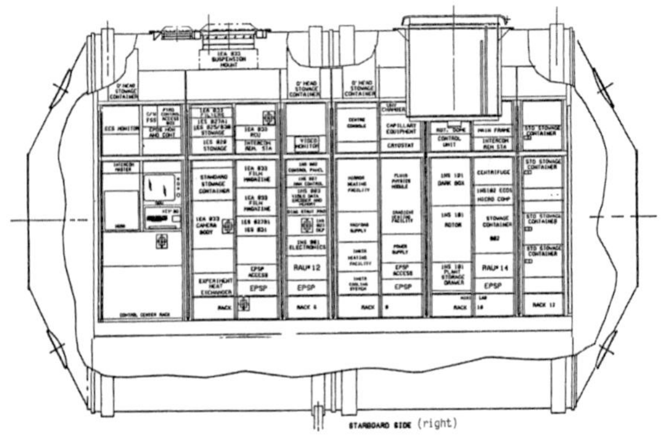

← FORWARD

ments: a pressurized, habitable laboratory called a module, in which scientists can work without cumbersome space suits; and unpressurized platforms called pallets designed to support instruments such as telescopes, sensors and antennas which require direct exposure to space. These elements may be used separately or in various combinations, returned to earth, and reused on other flights.

The module comes in two 4-meter (13.1-foot) diameter sections -- a core segment and an experiment segment. The core segment houses data processing equipment and utilities for the module and pallets when both are flown together. It also has laboratory fixtures such as air-cooled, standard 48.26-centimeter (19-inch) experiment racks, a work bench, and provision for accommodating a high-quality win-

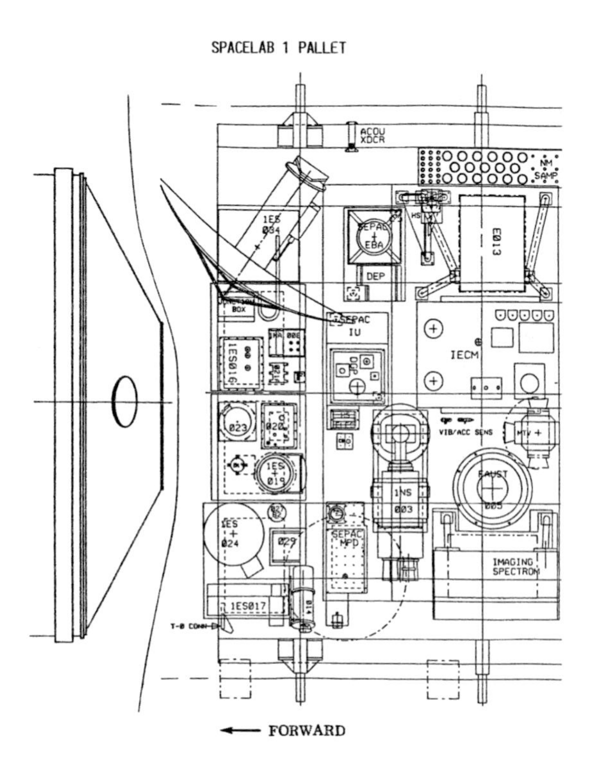

dow/viewport assembly for optical experiments and photography.

The second section, called the experiment segment, provides further pressurized work area, space for additional experiment racks, and provision for mounting either the window assembly or a scientific airlock designed to enable the crew to expose experiments carried in the module to the space environment. The core segment can be flown by itself, in what is called the short module configuration, or coupled in tandem with the experiment segment in the long module configuration.

The short module measures 4.26 m (15.4 ft.) in length overall and consists of the core segment and two cone-shaped end sections. The long module, including end cones, is 7 m (23 ft.) long. When the habitable module is flown, a 1-m (3.3-ft.) diameter enclosed passageway called the Spacelab Transfer Tunnel, connects the module with the mid-deck of the orbiter. The tunnel can be assembled in two lengths. A 2.7-m (8.7-ft.) tunnel can be used for missions during which the module is carried in the forward portion of the payload bay. A 5.8-m (18.8-ft.) tunnel can be used for missions on which the module is carried in the aft portion of the bay. Spacelab pallets are U-shaped aluminum frame and panel structures 4 m (13.1 ft.) wide and 3 m (10 ft.) long. Heavy equipment is mounted on the pallet frame using a series of connectors called hard points. Light weight equipment can be mounted directly onto the panels. As many as five Spacelab pallets can be flown in the cargo bay, individually or with two or three linked together to form trains. When pallets are flown without the module, the subsystems necessary for experiment operation are contained in a pressurized cylinder called an "igloo."

When no module is flown, payload specialists operate the experiments from the aft flight deck of the Shuttle orbiter. When the module is flown, the Spacelab subsystems are carried in the module.

SPACE SHUTTLE ASTRONAUTS

Acton,Loren—Jul 29-Aug 6/85 -Ch-51F
Adamson,James—Aug 2-11/91 -A-43, Aug 8-13/89 -Co-28
Akers,Thomas—Sep 16-26/96 -A-79, Oct 6-10/90 -D-41, Dec 2-13/93 -E-61, May 7-16/92 -E-49
Al Sa'ud,Sultan —Jun 17-24/85 -D-51G
Allen,Andrew—Jul 31-Aug 8/92 -A-46, Mar 4-18/94 -Co-62, Feb 22-Mar 9/96 -Co-75
Allen,Joseph—Nov 11-16/82 -Co-5, Nov 8-16/84 -D-51A
Altman,Scott—Sep 8-20/00 -A-106, Apr 17-May 3/98 -Co-90, Mar 1-12/02 -Co-109
Anderson,Clayton—Sep 7-?/07 -A-120, Jun 28-?/07 -E-118
Anderson,Michael—Jan 16-Feb 1/03 -Co-107, Jan 22-31/98 -E-89
Apt,Jay—Apr 9-20/94 -E-59
Apt,Jerome—Apr 5-11/91 -A-37, Sep 16-26/96 -A-79, Sep 12-20/92 -E-47
Archambault,Lee—Mar 16-?/07 -A-117
Ashby,Jeffrey—Jul 23-27/99 -Co-93, Apr 19-May 1/01 -E-100, Oct 7-18/02 -A-112
Bagian,James—Jun 5-14/91 -Co-40, Mar 13-18/89 -D-29
Baker,Ellen—Jun 27-Jul 7/95 -A-71, Oct 18-23/89 -A-34, Jun 25-Jul 9/92 -Co-50
Baker,Michael—Aug 2-11/91 -A-43, Oct 22-Nov 1/92 -Co-52, Jan 12-22/97 -A-81, Sep 30-Oct 11/94 -E-68
Barry,Daniel—Aug 10-22/01 -D-105, May 27-Jun 6/99 -D-96, Jan 11-20/96 -E-72
Bartoe,John-David—Jul 29-Aug 6/85 -Ch-51F
Baudry,Patrick—Jun 17-24/85 -D-51G
Blaha,John—Jan 12-22/97 -A-81, Sep 16-26/96 -A-79, Mar 13-18/89 -D-29, Nov 22-27/89 -D-33, Aug 2-11/91 -A-43, Oct 18-Nov 1/93 -Co-58
Bloomfield,Michael—Sep 25-Oct 6/97 -A-86, Nov 30-Dec 11/00 -E-97, Apr 8-19/02 -A-110
Bluford,Guion—Aug 30-Sep 5/83 -Ch-8, Oct 30-Nov 6/85 -Ch-61A, Apr 28-May 6/91 -D-39, Dec 2-9/92 -D-53
Bobko,Karol—Apr 4-9/83 -Ch-6, Oct 3-7/85 -A-51J, Apr 12-19/85 -D-51D
Bolden,Charles—Jan 12-18/86 -Co-61C, Apr 24-29/90 -D-31, Mar 24-Apr 2/92 -A-45, Feb 3-11/94 -D-60
Bondar,Roberta—Jan 22-30/92 -D-42
Bowersox,Kenneth—Jun 25-Jul 9/92 -Co-50, Dec 2-13/93 -E-61, Oct 20-Nov 5/95 -Co-73, Feb 11-21/97 -D-82
Brady,Charles—Jun 20-Jul 7/96 -Co-78
Brand,Vance—Feb 3-11/84 -Ch-41B, Dec 2-10/90 -Co-35, Nov 11-16/82 -Co-5
Brandenstein,Daniel—Aug 30-Sep 5/83 -Ch-8, Jan 9-20/90 -Co-32, Jun 17-24/85 -D-51G, May 7-16/92 -E-49
Bridges,Roy—Jul 29-Aug 6/85 -Ch-51F
Brown,Curtis—Nov 3-14/94 -A-66, May 19-29/96 -E-77, Sep 12-20/92 -E-47, Aug 7-19/97 -D-85, Dec 19-27/99 -D-103, Oct 29-Nov 7/98 -D-95
Brown,David—Jan 16-Feb 1/03 -Co-107
Brown,Mark—Aug 8-13/89 -Co-28, Sep 12-18/91 -D-48
Buchli,James—Oct 30-Nov 6/85 -Ch-61A, Jan 24-27/85 -D-51C, Mar 13-18/89 -D-29, Sep 12-18/91 -D-48
Buckey,Jay—Apr 17-May 3/98 -Co-90
Budarin,Nikolai—Jun 27-Jul 7/95 -A-71
Burbank,Daniel—Sep 8-20/00 -A-106, Sep 9-21/06 -A-115
Bursch,Daniel—Sep 12-22/93 -D-51, Dec 5-17/01 -E-108, May 19-29/96 -E-77, Sep 30-Oct 11/94 -E-68
Cabana,Robert—Dec 2-9/92 -D-53, Oct 6-10/90 -D-41, Jul 8-23/94 -Co-65, Dec 4-15/98 -E-88
Caldwell,Tracy—Jun 28-?/07 -E-118
Camarda,Charles—Jul 26-Aug 9/05 -D-114
Cameron,Kenneth—Apr 5-11/91 -A-37, Nov 12-20/95 -A-74, Apr 8-17/93 -D-56
Carey,Duane—Mar 1-12/02 -Co-109
Carter,Manley—Nov 22-27/89 -D-33
Casper,John—Feb 28-Mar 44/90 -A-36, Mar 4-18/94 -Co-62, Jan 13-19/93 -E-54, May 19-29/96 -E-77
Cenker,Robert—Jan 12-18/86 -Co-61C
Chang-Diaz,Franklin—Jul 31-Aug 8/92 -A-46, Oct 18-23/89 -A-34, Feb 22-Mar 9/96 -Co-75, Jan 12-18/86 -Co-61C, Feb 3-11/94 -D-60, Jun 2-12/98 -D-91, Jun 5-19/02 -E-111
Chawla,Kalpana—Jan 16-Feb 1/03 -Co-107, Nov 19-Dec 5/97 -Co-87
Cheli,Maurizio—Feb 22-Mar 9/96 -Co-75
Chiao,Leroy—Jul 8-23/94 -Co-65, Oct 11-24/00 -D-92, Jan 11-20/96 -E-72
Chilton,Kevin—Apr 9-20/94 -E-59, May 7-16/92 -E-49, Mar 22-31/96 -A-76
Chretien,Jean-Loup—Sep 25-Oct 6/97 -A-86
Clark,Laurel—Jan 16-Feb 1/03 -Co-107
Cleave,Mary—May 4-8/89 -A-30, Nov 26-Dec 3/85 -A-61B
Clervoy,Jean-Francis—Nov 3-14/94 -A-66, May 15-24/97 -A-84, Dec 19-27/99 -D-103
Clifford,Michael—Dec 2-9/92 -D-53, Mar 22-31/96 -A-76, Apr 9-20/94 -E-59
Coats,Michael—Aug 30-Sep 5/84 -D-41D, Apr 28-May 6/91 -D-39, Mar 13-18/89 -D-29
Cockrell,Kenneth—Apr 8-17/93 -D-56, Sep 7-18/95 -E-69, Feb 7-20/01 -A-98, Nov 19-Dec 7/96 -Co-80, Jun 5-19/02 -E-111
Coleman,Catherine—Jul 23-27/99 -Co-93, Oct 20-Nov 5/95 -Co-73
Collins,Eileen—May 15-24/97 -A-84, Feb 3-11/95 -D-63, Jul 23-27/99 -Co-93, Jul 26-Aug 9/05 -D-114
Covey,Richard—Aug 27-Sep 3/85 -D-51I, Sep 29-Oct 3/88 -D-26, Nov 15-20/90 -A-38, Dec 2-13/93 -E-61
Creighton,John—Jun 17-24/85 -D-51G, Feb 28-Mar 44/90 -A-36, Sep 12-18/91 -D-48
Crippen,Robert—Apr 12-14/81 -Co-1, Apr 6-13/84 -Ch-41C, Jun 18-24/83 -Ch-7, Oct 5-13/84 -Ch-41G

Crouch,Roger—Apr 4-8/97 -Co-83, Jul 1-17/97 -Co-94
Culbertson,Frank—Nov 15-20/90 -A-38, Aug 10-22/01 -D-105, Dec 5-17/01 -E-108, Sep 12-22/93 -D-51
Curbeam,Robert—Feb 7-20/01 -A-98, Aug 7-19/97 -D-85, Dec 7-?/06 -D-116
Currie,Nancy—Mar 1-12/02 -Co-109, Jul 13-22/95 -D-70, Dec 4-15/98 -E-88
David Jones,Thomas—Nov 19-Dec 7/96 -Co-80, Sep 30-Oct 11/94 -E-68
Davis,Jan—Aug 7-19/97 -D-85, Feb 3-11/94 -D-60, Sep 12-20/92 -E-47
DeLucas,Lawrence—Jun 25-Jul 9/92 -Co-50
Dezhurov,Vladimir—Jun 27-Jul 7/95 -A-71, Aug 10-22/01 -D-105, Dec 5-17/01 -E-108
Doi,Takao—Nov 19-Dec 5/97 -Co-87
Duffy,Brian—Mar 24-Apr 2/92 -A-45, Jun 21-Jul 1/93 -E-57, Oct 11-24/00 -D-92, Jan 11-20/96 -E-72
Dunbar,Bonnie—Jun 27-Jul 7/95 -A-71, Oct 30-Nov 6/85 -Ch-61A, Jan 9-20/90 -Co-32, Jun 25-Jul 9/92 -Co-50, Jan 22-31/98 -E-89
Duque,Pedro—Oct 29-Nov 7/98 -D-95
Durrance,Samuel—Dec 2-10/90 -Co-35, Mar 2-18/95 -E-67
Edwards,Joe Frank—Jan 22-31/98 -E-89
England,Anthony—Jul 29-Aug 6/85 -Ch-51F
Engle,Joe—Nov 12-14/81 -Co-2, Aug 27-Sep 3/85 -D-51I
Fabian,John—Jun 18-24/83 -Ch-7, Jun 17-24/85 -D-51G
Favier,Jean-Jacques—Jun 20-Jul 7/96 -Co-78
Ferguson,Christopher—Sep 9-21/06 -A-115
Fettman,Martin—Oct 18-Nov 1/93 -Co-58
Fisher,Anna—Nov 8-16/84 -D-51A
Fisher,William—Aug 27-Sep 3/85 -D-51I
Foale,Michael—Mar 24-Apr 2/92 -A-45, May 15-24/97 -A-84, Sep 25-Oct 6/97 -A-86, Apr 8-17/93 -D-56, Dec 19-27/99 -D-103, Feb 3-11/95 -D-63
Foreman,Mike—Sep 7-?/07 -A-120
Forrester,Patrick—Mar 16-?/07 -A-117, Aug 10-22/01 -D-105
Fossum,Michael—Jul 4-17/06 -D-121
Frick,Stephen—Apr 8-19/02 -A-110
Frimout,Dirk—Mar 24-Apr 2/92 -A-45
Fuglesang,Christer—Dec 7-?/06 -D-116
Fullerton,Gordon—Mar 22-30/82 -Co-3, Jul 29-Aug 6/85 -Ch-51F
Furrer,Reinhard—Oct 30-Nov 6/85 -Ch-61A
Gaffney,Drew—Jun 5-14/91 -Co-40
Gardner,Dale—Aug 30-Sep 5/83 -Ch-8, Nov 8-16/84 -D-51A
Gardner,Guy—Dec 2-6/88 -A-27, Dec 2-10/90 -Co-35
Garn,Jake—Apr 12-19/85 -D-51D
Garneau,Marc—Oct 5-13/84 -Ch-41G, May 19-29/96 -E-77, Nov 30-Dec 11/00 -E-97
Garriott,Owen—Nov 28-Dec 8/83 -Co-9
Gemar,Charles—Nov 15-20/90 -A-38, Mar 4-18/94 -Co-62, Sep 12-18/91 -D-48
Gernhardt,Michael—Jul 12-24/01 -A-104, Apr 4-8/97 -Co-83, Jul 1-17/97 -Co-94, Sep 7-18/95 -E-69
Gibson,Robert—Feb 3-11/84 -Ch-41B, Dec 2-6/88 -A-27, Jun 27-Jul 7/95 -A-71, Jan 12-18/86 -Co-61C, Sep 12-20/92 -E-47
Gidzenko,Yuri—Mar 8-21/01 -D-102
Glenn,John—Oct 29-Nov 7/98 -D-95
Godwin,Linda—Apr 5-11/91 -A-37, Mar 22-31/96 -A-76, Apr 9-20/94 -E-59, Dec 5-17/01 -E-108
Gorie,Dominic—Jun 2-12/98 -D-91, Feb 11-22/00 -E-99, Dec 5-17/01 -E-108
Grabe,Ronald—May 4-8/89 -A-30, Oct 3-7/85 -A-51J, Jan 22-30/92 -D-42, June 21-Jul 1/93 -E-57
Gregory,Frederick—Apr 29-May 6/85 -Ch-51B, Nov 24-Dec 1/91 -A-44, Nov 22-27/89 -D-33
Gregory,William—Mar 2-18/95 -E-67
Griggs,David—Apr 12-19/85 -D-51D
Grunsfeld,John—Jan 12-22/97 -A-81, Mar 1-12/02 -Co-109, Dec 19-27/99 -D-103, Mar 2-18/95 -E-67
Guidoni,Umberto—Feb 22-Mar 9/96 -Co-75, Apr 19-May 1/01 -E-100
Gutierrez,Sidney—Jun 5-14/91 -Co-40, Apr 9-20/94 -E-59
Hadfield,Chris—Nov 12-20/95 -A-74, Apr 19-May 1/01 -E-100
Halsell,James—Nov 12-20/95 -A-74, Jul 8-23/94 -Co-65, May 19-29/00 -A-101, Apr 4-8/97 -Co-83, Jul 1-17/97 -Co-94
Hammond,Blaine—Apr 28-May 6/91 -D-39, Sep 9-20/94 -D-64
Harbaugh,Gregory—Jun 27-Jul 7/95 -A-71, Apr 28-May 6/91 -D-39, Feb 11-21/97 -D-82, Jan 13-19/93 -E-54
Harris,Bernard—Apr 26-May 6/93 -Co-55, Feb 3-11/95 -D-63
Hart,Terry—Apr 6-13/84 -Ch-41C
Hartsfield,Henry—Jun 27-Jul 4/82 -Co-4, Oct 30-Nov 6/85 -Ch-61A, Aug 30-Sep 5/84 -D-41D
Hauck,Frederick—Jun 18-24/83 -Ch-7, Nov 8-16/84 -D-51A, Sep 29-Oct 3/88 -D-26
Hawley,Staven—Jan 12-18/86 -Co-61C, Jul 23-27/99 -Co-93, Apr 24-29/90 -D-31, Aug 30-Sep 5/84 -D-41D, Feb 11-21/97 -D-82
Helms,Susan—May 19-29/00 -A-101, Jun 20-Jul 7/96 -Co-78, Aug 10-22/01 -D-105, Mar 8-21/01 -D-102, Sep 9-20/94 -D-64, Jan 13-19/93 -E-54
Henize,Karl—Jul 29-Aug 6/85 -Ch-51F
Hennen,Thomas—Nov 24-Dec 1/91 -A-44
Henricks,Terrence—Nov 24-Dec 1/91 -A-44, Apr 26-May 6/93 -Co-55, Jun 20-Jul 7/96 -Co-78, Jul 13-22/95 -D-70
Herrington,John—Nov 23-Dec 7/02 -E-113
Hieb,Richard—Jul 8-23/94 -Co-65, Apr 28-May 6/91 -D-39, May 7-16/92 -E-49
Higginbotham,Joan—Dec 7-?/06 -D-116
Hilmers,Davd—Oct 3-7/85 -A-51J, Feb 28-Mar 44/90 -A-36, Jan 22-30/92 -D-42, Sep 29-Oct 3/88 -D-26
Hire,Kathryn—Apr 17-May 3/98 -Co-90
Hobaugh,Charles—Jul 12-24/01 -A-104, Jun 28-?/07 -E-118

Hoffman,Jeffrey—Jul 31-Aug 8/92 -A-46, Dec 2-10/90 -Co-35, Feb 22-Mar 9/96 -Co-75, Apr 12-19/85 -D-51D, Dec 2-13/93 -E-61
Horowitz,Scott—May 19-29/00 -A-101, Feb 22-Mar 9/96 -Co-75, Feb 11-21/97 -D-82, Aug 10-22/01 -D-105
Hughes-Fulford,Millie—Jun 5-14/91 -Co-40
Husband,Rick—May 27-Jun 6/99 -D-96, Jan 16-Feb 1/03 -Co-107
Ivins,Marsha—Feb 7-20/01 -A-98, Jan 12-22/97 -A-81, Jul 31-Aug 8/92 -A-46, Jan 9-20/90 -Co-32, Mar 4-18/94 -Co-62
Jarvis,Gregory—Jan 28/86 -Ch-51L
Jemison,Mae—Sep 12-20/92 -E-47
Jernigan,Tamara—Nov 19-Dec 7/96 -Co-80, Oct 22-Nov 1/92 -Co-52, May 27-Jun 6/99 -D-96, Mar 2-18/95 -E-67
Jett,Brent—Jan 12-22/97 -A-81, Jan 11-20/96 -E-72, Sep 9-21/06 -A-115, Nov 30-Dec 11/00 -E-97
Jones,Thomas—Feb 7-20/01 -A-98, Apr 9-20/94 -E-59
Kadenyuk,Leonid—Nov 19-Dec 5/97 -Co-87
Kavandi,Janet—Jul 12-24/01 -A-104, Jun 2-12/98 -D-91, Feb 11-22/00 -E-99
Kelly,James—Jul 26-Aug 9/05 -D-114, Mar 8-21/01 -D-102
Kelly,Mark—Jul 4-17/06 -D-121, Dec 5-17/01 -E-108
Kelly,Scott—Dec 19-27/99 -D-103, Jun 28-?/07 -E-118
Kondakova,Elena—May 15-24/97 -A-84
Kregel,Kevin—Jun 20-Jul 7/96 -Co-78, Jul 13-22/95 -D-70, Nov 19-Dec 5/97 -Co-87, Feb 11-22/00 -E-99
Krikalev,Sergei—Feb 3-11/94 -D-60, Mar 8-21/01 -D-102, Dec 4-15/98 -E-88
Lawrence,Wendy—Sep 25-Oct 6/97 -A-86, Jul 26-Aug 9/05 -D-114, Jun 2-12/98 -D-91, Mar 2-18/95 -E-67
Lee,Mark—May 4-8/89 -A-30, Feb 11-21/97 -D-82, Sep 9-20/94 -D-64, Sep 12-20/92 -E-47
Leestma,David—Mar 24-Apr 2/92 -A-45, Oct 5-13/84 -Ch-41G, Aug 8-13/89 -Co-28
Lenoir,William—Nov 11-16/82 -Co-5
Leslie,Fred—Oct 20-Nov 5/95 -Co-73
Lichtenberg,Byron—Mar 24-Apr 2/92 -A-45, Nov 28-Dec 8/83 -Co-9
Lind,Don Leslie—Apr 29-May 6/85 -Ch-51B
Lindsey,Steven—Nov 19-Dec 5/97 -Co-87, Oct 29-Nov 7/98 -D-95, Jul 12-24/01 -A-104, Jul 4-17/06 -D-121
Linenger,Jerry—Jan 12-22/97 -A-81, May 15-24/97 -A-84, Sep 9-20/94 -D-64
Linnehan,Richard—Apr 17-May 3/98 -Co-90, Jun 20-Jul 7/96 -Co-78, Mar 1-12/02 -Co-109
Linteris,Gregory—Apr 4-8/97 -Co-83, Jul 1-17/97 -Co-94
Lockhart,Paul—Jun 5-19/02 -E-111, Nov 23-Dec 7/02 -E-113
Lonchakov,Yuri—Apr 19-May 1/01 -E-100
Lopez-Alegria,Michael—Oct 20-Nov 5/95 -Co-73, Oct 11-24/00 -D-92, Nov 23-Dec 7/02 -E-113
Lounge,John—Dec 2-10/90 -Co-35, Aug 27-Sep 3/85 -D-51I, Sep 29-Oct 3/88 -D-26
Lousma,Jack—Mar 22-30/82 -Co-3
Low,David—Aug 2-11/91 -A-43, Jan 9-20/90 -Co-32, June 21-Jul 1/93 -E-57
Lu,Edward—May 15-24/97 -A-84, Sep 8-20/00 -A-106
Lucid,Shannon—Aug 2-11/91 -A-43, Mar 22-31/96 -A-76, Oct 18-23/89 -A-34, Sep 16-26/96 -A-79, Oct 18-Nov 1/93 -Co-58, Jun 17-24/85 -D-51G
MacLean,Steven—Sep 9-21/06 -A-115, Oct 22-Nov 1/92 -Co-52
Magnus,Sandra—Oct 7-18/02 -A-112
Malenchenko,Yuri—Sep 8-20/00 -A-106
Malerbo,Franco—Jul 31-Aug 8/92 -A-46
Massimino,Michael—Mar 1-12/02 -Co-109
Mastracchio,Richard—Sep 8-20/00 -A-106, Jun 28-?/07 -E-118
Mattingly,Ken—Jun 27-Jul 4/82 -Co-4, Jan 24-27/85 -D-51C
McArthur,William—Nov 12-20/95 -A-74, Oct 18-Nov 1/93 -Co-58, Oct 11-24/00 -D-92
McAuliffe,Christa—Jan 28/86 -Ch-51L
McBride,Jon—Oct 5-13/84 -Ch-41G
McCandless,Bruce—Feb 3-11/84 -Ch-41B, Apr 24-29/90 -D-31
McCool,William—Jan 16-Feb 1/03 -Co-107
McCulley,Michael—Oct 18-23/89 -A-34
McMonagle,Donald—Apr 28-May 6/91 -D-39, Jan 13-19/93 -E-54, Nov 3-14/94 -A-66
McNair,Ronald—Feb 3-11/84 -Ch-41B, Jan 28/86 -Ch-51L
Meade,Carl—Nov 15-20/90 -A-38, Jun 25-Jul 9/92 -Co-50, Sep 9-20/94 -D-64
Melnick,Bruce—Oct 6-10/90 -D-41, May 7-16/92 -E-49
Melroy,Pamela—Oct 11-24/00 -D-92, Sep 7-?/07 -A-120, Oct 7-18/02 -A-112
Merbold,Ulf—Nov 28-Dec 8/83 -Co-9, Jan 22-30/92 -D-42
Messerschmid,Ernst—Oct 30-Nov 6/85 -Ch-61A
Mohri,Mamoru—Feb 11-22/00 -E-99, Sep 12-20/92 -E-47
Morgan,Barbara—Jun 28-?/07 -E-118
Morin,Lee—Apr 8-19/02 -A-110
Morukov,Boris—Sep 8-20/00 -A-106
Mukai,Chiaki—Jul 8-23/94 -Co-65, Oct 29-Nov 7/98 -D-95
Mullane,Richard—Dec 2-6/88 -A-27, Feb 28-Mar 44/90 -A-36, Aug 30-Sep 5/84 -D-41D
Musgrave,Story—Nov 24-Dec 1/91 -A-44, Apr 4-9/83 -Ch-6, Jul 29-Aug 6/85 -Ch-51F, Nov 19-Dec 7/96 -Co-80, Nov 22-27/89 -D-33, Dec 2-13/93 -E-61
Nagel,Steven—Oct 30-Nov 6/85 -Ch-61A, Jun 17-24/85 -D-51G, Apr 5-11/91 -A-37, Apr 26-May 6/93 -Co-55
Nelson,Bill—Jan 12-18/86 -Co-61C
Nelson,George—Apr 6-13/84 -Ch-41C, Jan 12-18/86 -Co-61C, Sep 29-Oct 3/88 -D-26

Thagard,Norman—Jun 27-Jul 7/95 -A-71, May 4-8/89 -A-30, Apr 29-May 6/85 -Ch-51B, Jun 18-24/83 -Ch-7, Jan 22-30/92 -D-42
Thiele,Gerhard—Feb 11-22/00 -E-99
Thirsk,Robert Brent—Jun 20-Jul 7/96 -Co-78
Thomas,Andrew—Jul 26-Aug 9/05 -D-114, Jun 2-12/98 -D-91, Mar 8-21/01 -D-102, Jan 22-31/98 -E-89, May 19-29/96 -E-77
Thomas,Donald—Apr 4-8/97 -Co-83, Jul 1-17/97 -Co-94, Jul 8-23/94 -Co-65, Jul 13-22/95 -D-70
Thornton,Kathryn—Oct 20-Nov 5/95 -Co-73, Nov 22-27/89 -D-33, Dec 2-13/93 -E-61, May 7-16/92 -E-49
Thornton,William—Apr 29-May 6/85 -Ch-51B, Aug 30-Sep 5/83 -Ch-8
Thuot,Pierre—Feb 28-Mar 44/90 -A-36, Mar 4-18/94 -Co-62, May 7-16/92 -E-49
Titov,Vladimir—Sep 25-Oct 6/97 -A-86, Feb 3-11/95 -D-63
Tognini,Michael—Jul 23-27/99 -Co-93
Tokarev,Valery—May 27-Jun 6/99 -D-96
Trinh,Eugen—Jun 25-Jul 9/92 -Co-50
Truly,Richard—Nov 12-14/81 -Co-2, Aug 30-Sep 5/83 -Ch-8
Tryggvason,Bjarni—Aug 7-19/97 -D-85
Turin,Mikhail—Aug 10-22/01 -D-105, Dec 5-17/01 -E-108
Usachev,Yuri—May 19-29/00 -A-101, Aug 10-22/01 -D-105, Mar 8-21/01 -D-102
van den Berg,Lodewijk—Apr 29-May 6/85 -Ch-51B
van Hoften,James—Apr 6-13/84 -Ch-41C, Aug 27-Sep 3/85 -D-51I
Veach,Charles—Apr 28-May 6/91 -D-39, Oct 22-Nov 1/92 -Co-52
Voss,James—May 19-29/00 -A-101, Nov 24-Dec 1/91 -A-44, Aug 10-22/01 -D-105, Dec 2-9/92 -D-53, Mar 8-21/01 -D-102, Sep 7-18/95 -E-69
Voss,Janice—Apr 4-8/97 -Co-83, Jul 1-17/97 -Co-94, Feb 3-11/95 -D-63, Feb 11-22/00 -E-99, June 21-Jul 1/93 -E-57
Wakata,Koichi—Oct 11-24/00 -D-92, Jan 11-20/96 -E-72
Walheim,Rex—Apr 8-19/02 -A-110
Walker,Charles—Nov 26-Dec 3/85 -A-61B, Apr 12-19/85 -D-51D, Aug 30-Sep 5/84 -D-41D
Walker,David—Nov 8-16/84 -D-51A, May 4-8/89 -A-30, Dec 2-9/92 -D-53, Sep 7-18/95 -E-69
Walter,Ulrich—Apr 26-May 6/93 -Co-55
Walz,Carl—Sep 16-26/96 -A-79, Jul 8-23/94 -Co-65, Sep 12-22/93 -D-51, Dec 5-17/01 -E-108
Wang,Taylor—Apr 29-May 6/85 -Ch-51B
Weber,Mary Ellen—May 19-29/00 -A-101, Jul 13-22/95 -D-70
Weitz,Paul—Apr 4-9/83 -Ch-6
Wetherbee,James—Jan 9-20/90 -Co-32, Sep 25-Oct 6/97 -A-86, Oct 22-Nov 1/92 -Co-52, Feb 3-11/95 -D-63, Mar 8-21/01 -D-102, Nov 23-Dec 7/02 -E-113
Wheelock,Doug—Sep 7-?/07 -A-120
Wilcutt,Terrence—Sep 16-26/96 -A-79, Sep 30-Oct 11/94 -E-68, Sep 8-20/00 -A-106, Jan 22-31/98 -E-89
Williams,Dave—Apr 17-May 3/98 -Co-90, Jun 28-?/07 -E-118
Williams,Donald—Apr 12-19/85 -D-51D, Oct 18-23/89 -A-34
Williams,Jeffrey—May 19-29/00 -A-101
Williams,Sunita—Dec 7-?/06 -D-116
Wilson,Stephanie—Jul 4-17/06 -D-121
Wisoff,Peter—Jan 12-22/97 -A-81, Oct 11-24/00 -D-92, Jun 21-Jul 1/93 -E-57, Sep 30-Oct 11/94 -E-68
Wolf,David—Oct 7-18/02 -A-112, Sep 25-Oct 6/97 -A-86, Oct 18-Nov 1/93 -Co-58, Jan 22-31/98 -E-89
Young,John—Apr 12-14/81 -Co-1, Nov 28-Dec 8/83 -Co-9
Yurchikhin,Fyodor—Oct 7-18/02 -A-112
Zamka,George—Sep 7-?/07 -A-120

SPACEHAB

STS-57 – Spacehab-1
STS-60 – Spacehab-2
STS-63 – Spacehab-3
STS-76 – Spacehab/MIR
STS-77 – Spacehab-4
STS-79 – Spacehab/MIR
STS-81 – Spacehab-6/MIR
STS-84 – Spacehab-7/MIR
STS-86 – Spacehab/MIR
STS-89 – Spacehab-8/MIR
STS-91 – Spacehab-9/MIR
STS-95 – Spacehab
STS-96 – Spacehab
STS-101 – Spacehab
STS-106 – Spacehab
STS-107 – Spacehab RDM
STS-116 – Spacehab/ICC

SPACE SHUTTLE SATELLITE LAUNCHES 1982-2006

ACTS	STS-51	SPARTAN-201	STS-64
ARABSAT-1B	STS-51G	SPARTAN-201	STS-69
ASC-1,	STS-51I	SPARTAN-201-04	STS-87
AUSSAT-1	STS-51I	SPARTAN-201-05	STS-95
AUSSAT-2	STS-61B	SPARTAN-204	STS-63
BREMSAT	STS-60	SPARTAN-207	STS-77
CHANDRA X-RAY OBS.	STS-93	SPAS	STS-66
DSP	STS-44	SPAS	STS-85
ERBS	STS-41G	SPAS	STS-41B
EURECA-1	STS-46	SPAS	STS-7
GALILEO	STS-34	STARSHINE	STS-96
GAMMA RAY OBS.	STS-37	STARSHINE-2	STS-108
GLOMR	STS-51B	SYNCOM-IV-1	STS-51A
HUBBLE SPACE TEL.	STS-31	SYNCOM-IV-2	STS-41D
INSAT	STS-8	SYNCOM-IV-3	STS-51D
LAGEOS II	STS-52	SYNCOM-IV-4	STS-51I
LDEF	STS-41C	SYNCOM-IV-5	STS-32
MAGELLAN	STS-30	TDRS-A	STS-6
MIGHTYSAT-1	STS-88	TDRS-C	STS-26
MORELOS-A	STS-51G	TDRS-D	STS-29
MORELOS-B	STS-61B	TDRS-E	STS-43
NUSAT	STS-51B	TDRS-F	STS-54
OAST	STS-72	TDRS-G	STS-70
ORFEUS-SPAS	STS-51	TELESAT-E	STS-5
ORFEUS-SPAS	STS-80	TELESAT-F	STS-7
PALAPA-B1	STS-7	TELESAT-H	STS-51A
PALAPA-B2	STS-41B	TELESAT-I	STS-51D
PAMS	STS-77	TELSTAR 3C	STS-41D
SAC-A	STS-88	TELSTAR 3D	STS-51G
SATCOM-KU-1	STS-61C	TSS-1	STS-46
SATCOM-KU-2	STS-61B	UARS	STS-48
SBS-C	STS-5	ULYSSES	STS-41
SBS-D	STS-41D	WESTAR	STS-41B
SIMPLESAT	STS-105	WSF	STS-80

MPLM

STS-102 – Leonardo
STS-100 – Rafaello
STS-105 – Leonardo
STS-108 – Rafaello
STS-111 – Leonardo
STS-121 – Leonardo

Multi-Purpose Logistics Module

SPACELAB

STS-9, Spacelab 1,Module (LM1) and Pallet
STS-51-B, Spacelab 3, Module (LM1)
STS-51-F, Spacelab 2, 3 Pallet configuration
STS-61-A, Spacelab D1, Module (LM2)
STS-35, ASTRO-1,Pallet
STS-40, SLS-1, Module (LM1)
STS-42, IML-1, Module (LM2)
STS-45, ATLAS-1, two Pallet configuration
STS-50, USML-1, Module (LM1)
STS-47, Spacelab-J, Module (LM2)
STS-56, ATLAS-2, Pallet
STS-55, Spacelab D2, Module (LM1)
STS-58, SLS-2, Module (LM2)
STS-59, SRL-1, Pallet
STS-65, IML-2, Module (LM1)
STS-68, SRL-2, Pallet
STS-66, ATLAS-3, Pallet
STS-67, ASTRO-2, Pallet
STS-71, Spacelab-Mir, Module (LM2)
STS-73, USML-2, Module (LM1)
STS-78, LMS, Module (LM2)
STS-83, MSL-1, Module (LM1)
STS-94, MSL-1R, Module (LM1)
STS-90, Neurolab, Module (LM2)
STS-99, SRTM, Pallet

MIR FLIGHTS

STS-63, STS-71, STS-74, STS-79, STS-81, STS-86, STS-91

ISS FLIGHTS (1998-2006)

STS-88	Unity Pressurized Mating Adapter
STS-92	Z1 truss
STS-97	P6 truss
STS-98	Destiny US Laboratory
STS-101	Construction & Maintenance
STS-106	Construction & Maintenance
STS-102	Construction & Maintenance
STS-100	SSRMS
STS-104	US Joint Airlock
STS-105	Construction & Maintenance
STS-108	Construction & Maintenance
STS-110	S0 Truss
STS-111	Mobile Remote
STS-112	S1 truss
STS-113	P1 truss
STS-115	P3/P4 truss
STS-121	Construction & Maintenance
STS-116	P6 arrays

GROUND TURNAROUND SEQUENCE

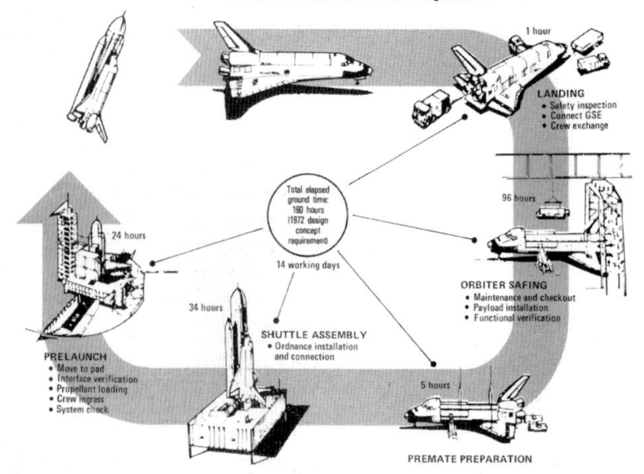

Legend for color section

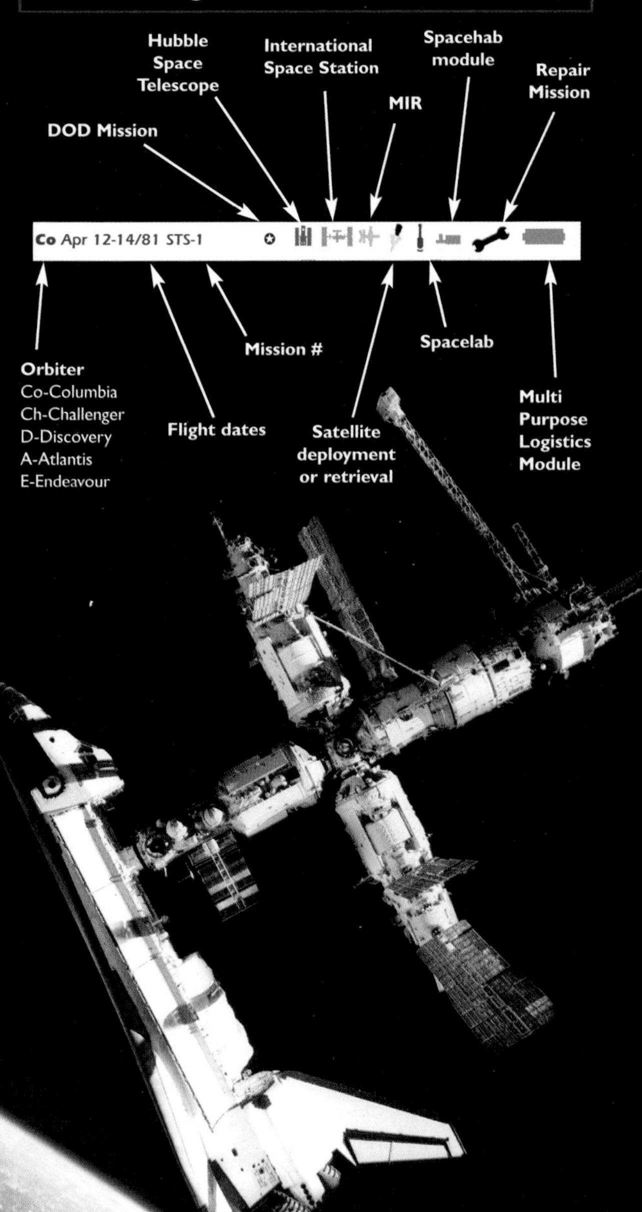

John Young (C)
Robert Crippen

Co Apr 12-14/81 STS-1

Joe Engle(C)
Richard Truly

Co Nov 12-14/81 STS-2

Jack Lousma (C)
Gordon Fullerton

Co Mar 22-30/82 STS-3

Ken Mattingly (C)
Henry Hartsfield

Co Jun 27-Jul 4/82 STS-4

Vance Brand (C)
Robert Overmyer
Joseph Allen
William Lenoir

Co Nov 11-16/82 STS-5

Paul Weitz (C)
Karol Bobko
Donald Peterson
Story Musgrave

Ch Apr 4-9/83 STS-6

Robert Crippen (C)
Frederick Hauck
Sally Ride
John Fabian
Norman Thagard

Ch Jun 18-24/83 STS-7

Richard Truly (C)
Daniel Brandenstein
Guion Bluford
Dale Gardner
William Thornton

Ch Aug 30-Sep 5/83 STS-8

John Young (C)
Brewster Shaw
Robert Parker
Owen Garriott
Byron Lichtenberg
Ulf Merbold

Co Nov 28-Dec 8/83 STS-9

Vance Brand (C)
Robert Gibson
Bruce McCandless
Robert Stewart
Ronald McNair

Ch Feb 3-11/84 STS-41B

Robert Crippen (C)
Francis Scobee
Terry Hart
James van Hoften
George Nelson

Ch Apr 6-13/84 STS-41C

Henry Hartsfield (C)
Michael Coats
Richard Mullane
Steven Hawley
Judith Resnick
Charles Walker

D Aug 30-Sep 5/84 STS-41D

Robert Crippen (C)
Jon McBride
David Leestma
Sally Ride
Kathryn Sullivan
Paul Scully-Power
Marc Garneau

Ch Oct 5-13/84 STS-41G

Frederick Hauck (C)
David Walker
Joseph Allen
Anna Fisher
Dale Gardner

D Nov 8-16/84 STS-51A

Ken Mattingly (C)
Loren Shriver
Ellison Onizuka
James Buchli
Gary Payton

D Jan 24-27/85 STS-51C

Karol Bobko (C)
Donald Williams
Jeffrey Hoffman
David Griggs
Rhea Seddon
Charles Walker
Jake Garn

D Apr 12-19/85 STS-51D

Robert Overmyer (C)
Frederick Gregory
Don Leslie Lind
Norman Thagard
William Thornton
Taylor Wang
Lodewijk van den Berg

Ch Apr 29-May 6/85 STS-51B

Daniel Brandenstein (C)
John Creighton
John Fabian
Steven Nagel
Shannon Lucid
Patrick Baudry
Sultan Salman Abdul
Azziz Al Sa'ud

D Jun 17-24/85 STS-51G

Gordon Fullerton (C)
Roy Bridges
Story Musgrave
Anthony England
Karl Henize
Loren Acton
John-David Bartoe

Ch Jul 29-Aug 6/85 STS-51F

Joe Engle (C)
Richard Covey
James van Hoften
William Fisher
John Lounge

D Aug 27-Sep 3/85 STS-51I

Karol Bobko (C)
Ronald Grabe
Davd Hilmers
Robert Stewart
William Pailes

A Oct 3-7/85 STS-51J

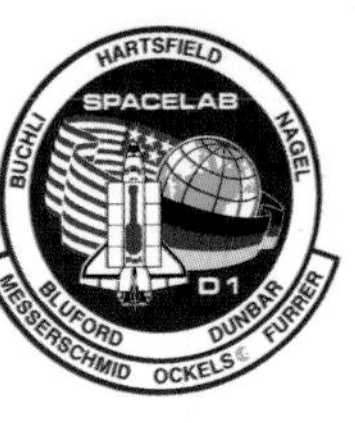

Henry Hartsfield (C)
Steven Nagel
James Buchli
Guion Bluford
Bonnie Dunbar
Reinhard Furrer
Wubbo Ockels
Ernst Messerschmid

Ch Oct 30-Nov 6/85 STS-61A

Brewster Shaw (C)
Bryan O'Connor
Sherwood Spring
Mary Cleave
Jerry Ross
Charles Walker
Rodolfo Neri Vela

A Nov 26-Dec 3/85 STS-61B

Robert Gibson (C)
Charles Bolden
George Nelson
Staven Hawley
Franklin Chang-Diaz
Robert Cenker
Bill Nelson

Co Jan 12-18/86 STS-61C

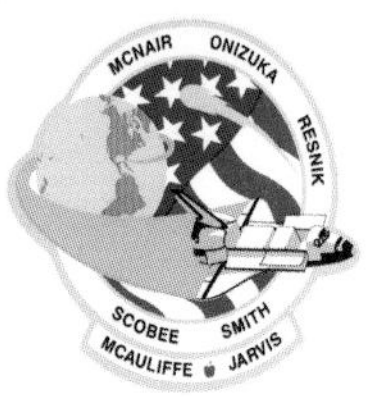

Francis Scobee (C)
Michael Smith
Ellison Onizuka
Judith Resnick
Ronald McNair
Gregory Jarvis
Christa McAuliffe

Ch Jan 28/86 STS-51L

Frederick Hauck (C)
Richard Covey
John Lounge
George Nelson
David Hilmers

D Sep 29-Oct 3/88 STS-26

Robert Gibson (C)
Guy Gardner
Richard Mullane
Jerry Ross
William Shepherd

A Dec 2-6/88 STS-27 ✪

Michael Coats (C)
John Blaha
James Buchli
Robert Springer
James Bagian

D Mar 13-18/89 STS-29

David Walker (C)
Ronald Grabe
Norman Thagard
Mary Cleave
Mark Lee

A May 4-8/89 STS-30

Brewster Shaw (C)
Richard Richards
David Leestma
James Adamson
Mark Brown

Co Aug 8-13/89 STS-28 ✪

Donald Williams (C)
Michael McCulley
Shannon Lucid
Ellen Baker
Franklin Chang-Diaz

A Oct 18-23/89 STS-34

Frederick Gregory (C)
John Blaha
Story Musgrave
Kathryn Thornton
Manley Carter

D Nov 22-27/89 STS-33

Daniel Brandenstein (C)
James Wetherbee
Bonnie Dunbar
David Low
Marsha Ivins

Co Jan 9-20/90 STS-32

John Creighton (C)
John Casper
David Hilmers
Richard Mullane
Pierre Thuot

A Feb 28-Mar 44/90 STS-36 ✪

Loren Shriver (C)
Charles Bolden
Steven Hawley
Bruce McCandless
Kathryn Sullivan

D Apr 24-29/90 STS-31

Richard Richards (C)
Robert Cabana
Bruce Melnick
William Shepherd
Thomas Akers

D Oct 6-10/90 STS-41

Richard Covey (C)
Frank Culbertson
Robert Springer
Carl Meade
Charles Gemar

A Nov 15-20/90 STS-38

Vance Brand (C)
Guy Gardner
Jeffrey Hoffman
John Lounge
Robert Parker
Samuel Durrance
Ronald Parise

Co Dec 2-10/90 STS-35

Steven Nagel (C)
Kenneth Cameron
Jerry Ross
Jerome Apt
Linda Godwin

A Apr 5-11/91 STS-37

Michael Coats (C)
Blaine Hammond
Gregory Harbaugh
Donald McMonagle
Guion Bluford
Charles Veach
Richard Hieb

D Apr 28-May 6/91 STS-39 ✪

Bryan O'Connor (C)
Sidney Gutierrez
Rhea Seddon
James Bagian
Drew Gaffney
Millie Hughes-Fulford

Co Jun 5-14/91 STS-40

John Blaha (C)
Michael Baker
Shannon Lucid
David Low
James Adamson

A Aug 2-11/91 STS-43

John Creighton (C)
Kenneth Reightler
James Buchli
Mark Brown
Charles Gemar

D Sep 12-18/91 STS-48

Frederick Gregory (C)
Terrence Henricks
Story Musgrave
Mario Runco
James Voss
Thomas Hennen

A Nov 24-Dec 1/91 STS-44

Ronald Grabe (C)
Stephen Oswald
David Hilmers
Norman Thagard
William Readdy
Ulf Merbold
Roberta Bondar

D Jan 22-30/92 STS-42

Charles Bolden (C)
Brian Duffy
Kathryn Sullivan
David Leestma
Michael Foale
Dirk Frimout
Byron Lichtenberg

A Mar 24-Apr 2/92 STS-45

Daniel Brandenstein (C)
Kevin Chilton
Pierre Thuot
Kathryn Thornton
Richard Hieb
Thomas Akers
Bruce Melnick

E May 7-16/92 STS-49

Richard Richards (C)
Kenneth Bowersox
Bonnie Dunbar
Ellen Baker
Carl Meade
Lawrence DeLucas
Eugen Trinh

Co Jun 25-Jul 9/92 STS-50

Loren Shriver (C)
Andrew Allen
Jeffrey Hoffman
Franklin Chang-Diaz
Claude Nicollier
Marsha Ivins
Franco Malerbo

A Jul 31-Aug 8/92 STS-46

Robert Gibson (C)
Curtis Brown
Mark Lee
Jerome Apt
Jan Davis
Mae Jemison
Mamoru Mohri

E Sep 12-20/92 STS-47

James Wetherbee (C)
Michael Baker
William Shepherd
Tamara Jernigan
Charles Veach
Steven MacLean

Co Oct 22-Nov 1/92 STS-52

David Walker (C)
Robert Cabana
Guion Bluford
James Voss
Michael Clifford

D Dec 2-9/92 STS-53 ✪

John Casper (C)
Donald McMonagle
Mario Runco
Gregory Harbaugh
Susan Helms

E Jan 13-19/93 STS-54

Kenneth Cameron (C)
Stephen Oswald
Michael Foale
Kenneth Cockrell
Ellen Ochoa

D Apr 8-17/93 STS-56

Steven Nagel (C)
Terrence Henricks
Jerry Ross
Charles Precourt
Bernard Harris
Ulrich Walter
Hans Schliegel

Co Apr 26-May 6/93 STS-55

Ronald Grabe (C)
Brian Duffy
David Low
Nancy Sherlock
Peter Wisoff
Janice Voss

E Jun 21-Jul 1/93 STS-57

Frank Culbertson (C)
William Readdy
James Newman
Daniel Bursch
Carl Walz

D Sep 12-22/93 STS-51

John Blaha (C)
Richard Searfoss
Rhea Seddon
Shannon Lucid
David Wolf
William McArthur
Martin Fettman

Co Oct 18-Nov 1/93 STS-58

Richard Covey (C)
Kenneth Bowersox
Story Musgrave
Thomas Akers
Jeffrey Hoffman
Kathryn Thornton
Claude Nicollier

E Dec 2-13/93 STS-61 

Charles Bolden (C)
Kenneth Reightler
Franklin Chang-Diaz
Jan Davis
Ronald Sega
Sergei Krikalev

D Feb 3-11/94 STS-60

John Casper (C)
Andrew Allen
Pierre Thuot
Charles Gemar
Marsha Ivins

Co Mar 4-18/94 STS-62

Sidney Gutierrez (C)
Kevin Chilton
Linda Godwin
Jay Apt
Michael Clifford
Thomas Jones

E Apr 9-20/94 STS-59

Robert Cabana (C)
James Halsell
Richard Hieb
Carl Walz
Leroy Chiao
Donald Thomas
Chiaki Mukai

Co Jul 8-23/94 STS-65

Richard Richards (C)
Blaine Hammond
Carl Meade
Mark Lee
Susan Helms
Jerry Linenger

Sep 9-20/94 STS-64

Michael Baker (C)
Terrence Wilcutt
Thomas David Jones
Steven Smith
Peter Wisoff
Daniel Bursch

Sep 30-Oct 11/94 STS-68

Donald McMonagle (C)
Curtis Brown
Ellen Ochoa
Scott Parazynski
Joseph Tanner
Jean-Francis Clervoy

Nov 3-14/94 STS-66

James Wetherbee (C)
Eileen Collins
Michael Foale
Janice Voss
Bernard Harris
Vladimir Titov

D Feb 3-11/95 STS-63

Stephen Oswald (C)
William Gregory
Tamara Jernigan
John Grunsfeld
Wendy Lawrence
Ronald Parise
Samuel Durrance

E Mar 2-18/95 STS-67

Robert Gibson (C)
Charles Precourt
Ellen Baker
Gregory Harbaugh
Bonnie Dunbar
Anatoly Solovyez
Nikolai Budarin
Vladimir Dezhurov
Gennadiy Strekalov
Norman Thagard

A Jun 27-Jul 7/95 STS-71

HENRICKS
KREGEL
THOMAS
CURRIE
WEBER
STS-70

Terrence Henricks (C)
Kevin Kregel
Nancy Currie
Donald Thomas
Mary Ellen Weber

D Jul 13-22/95 STS-70

David Walker (C)
Kenneth Cockrell
James Voss
James Newman
Michael Gernhardt

E Sep 7-18/95 STS-69

Kenneth Bowersox (C)
Kent Rominger
Kathryn Thornton
Catherine Coleman
Michael Lopez-Alegria
Fred Leslie
Albert Sacco

Co Oct 20-Nov 5/95 STS-73

Kenneth Cameron (C)
James Halsell
Chris Hadfield
Jerry Ross
William McArthur

A Nov 12-20/95 STS-74

Brian Duffy (C)
Brent Jett
Leroy Chiao
Winston Scott
Koichi Wakata
Daniel Barry

E Jan 11-20/96 STS-72

Andrew Allen (C)
Scott Horowitz
Franklin Chang-Diaz
Jeffrey Hoffman
Maurizio Cheli
Claude Nicollier
Umberto Guidoni

Co Feb 22-Mar 9/96 STS-75

Kevin Chilton (C)
Richard Searfoss
Shannon Lucid
Linda Godwin
Michael Clifford
Ronald Sega

A Mar 22-31/96 STS-76

John Casper (C)
Curtis Brown
Daniel Bursch
Andrew Thomas
Marc Garneau
Mario Runco

E May 19-29/96 STS-77

Terrence Henricks (C)
Kevin Kregel
Susan Helms
Richard Linnehan
Charles Brady
Jean-Jacques Favier
Robert Brent Thirsk

Co Jun 20-Jul 7/96 STS-78

William Readdy (C)
Terrence Wilcutt
Thomas Akers
Jerome Apt
Carl Walz
John Blaha
Shannon Lucid

A Sep 16-26/96 STS-79

Kenneth Cockrell (C)
Kent Rominger
Tamara Jernigan
Thomas David Jones
Story Musgrave

Co Nov 19-Dec 7/96 STS-80

Michael Baker (C)
Brent Jett
John Grunsfeld
Marsha Ivins
Peter Wisoff
Jerry Linenger
John Blaha

A Jan 12-22/97 STS-81

Kenneth Bowersox (C)
Scott Horowitz
Mark Lee
Gregory Harbaugh
Steven Smith
Joseph Tanner
Steven Hawley

D Feb 11-21/97 STS-82

James Halsell (C)
Susan Still
Janice Voss
Donald Thomas
Michael Gernhardt
Roger Crouch
Gregory Linteris

Co Apr 4-8/97 STS-83

Charles Precourt (C)
Eileen Collins
Jerry Linenger
Michael Foale
Elena Kondakova
Carlos Noriega
Edward Lu
Jean-Francois Clervoy

A May 15-24/97 STS-84

James Halsell (C)
Susan Still
Janice Voss
Donald Thomas
Michael Gernhardt
Roger Crouch
Gregory Linteris

Co Jul 1-17/97 STS-94

Curtis Brown (C)
Kent Rominger
Jan Davis
Robert Curbeam
Stephen Robinson
Bjarni Tryggvason

D Aug 7-19/97 STS-85

James Wetherbee (C)
Michael Bloomfield
Vladimir Titov
Scott Parazynski
Jean-Loup Chretien
Wendy Lawrence
David Wolf
Michael Foale

A Sep 25-Oct 6/97 STS-86

Kevin Kregel (C)
Steven Lindsey
Kalpana Chawla
Winston Scott
Takao Doi
Leonid Kadenyuk

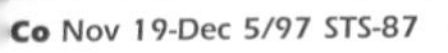

Co Nov 19-Dec 5/97 STS-87

Terrence Wilcutt (C)
Joe Frank Edwards
James Reilly
Michael Anderson
Bonnie Dunbar
Salizhan Sharipov
Andrew Thomas
David Wolf

E Jan 22-31/98 STS-89

Richard Searfoss (C)
Scott Altman
Kathryn Hire
Richard Linnehan
Dave Williams
Jay Buckey
James Pawelczyk

Co Apr 17-May 3/98 STS-90

Charles Precourt (C)
Dominic Gorie
Franklin Chang-Diaz
Wendy Lawrence
Janet Kavandi
Valeriy Ryumin
Andrew Thomas

D Jun 2-12/98 STS-91

Curtis Brown (C)
Steven Lindsey
Scott Parazynski
Stephen Robinson
Pedro Duque
Chiaki Mukai
John Glenn

D Oct 29-Nov 7/98 STS-95

Robert Cabana (C)
Frederick Sturckow
Nancy Currie
Jerry Ross
James Newman
Sergei Krikalev

E Dec 4-15/98 STS-88

Kent Rominger (C)
Rick Husband
Ellen Ochoa
Tamara Jernigan
Daniel Barry
Julie Payette
Valery Tokarev

D May 27-Jun 6/99 STS-96

Eileen Collins (C)
Jeffrey Ashby
Steven Hawley
Catherine Coleman
Michael Tognini

Co Jul 23-27/99 STS-93

Curtis Brown (C)
Scott Kelly
Jean-Francois Clervoy
Steven Smith
Michael Foale
John Grunsfeld
Claude Nicollier

D Dec 19-27/99 STS-103

Kevin Kregel (C)
Dom Gorie
Gerhard Thiele
Janet Kavandi
Janice Voss
Mamoru Mohri

E Feb 11-22/00 STS-99

James Halsell (C)
Scott Horowitz
Mary Ellen Weber
Jeffrey Williams
James Voss
Susan Helms
Yuri Usachev

A May 19-29/00 STS-101

Terrence Wilcutt (C)
Scott Altman
Edward Lu
Richard Mastracchio
Daniel Burbank
Yuri Malenchenko
Boris Morukov

A Sep 8-20/00 STS-106

Brian Duffy (C)
Pamela Melroy
Koichi Wakata
Peter Wisoff
Leroy Chiao
William McArthur
Michael Lopez-Alegria

D Oct 11-24/00 STS-92

Brent Jett (C)
Michael Bloomfield
Marc Garneau
Joseph Tanner
Carlos Noriega

E Nov 30-Dec 11/00 STS-97

Kenneth Cockrell (C)
Mark Polansky
Marsha Ivins
Thomas Jones
Robert Curbeam

A Feb 7-20/01 STS-98

James Wetherbee (C)
James Kelly
Andrew Thomas
Paul Richards
James Voss
Susan Helms
Yuri Usachev
William Shepherd
Yuri Gidzenko
Sergei Krikalev

D Mar 8-21/01 STS-102

Kent Rominger (C)
Jeffrey Ashby
Chris Hadfield
John Phillips
Scott Parazynski
Umberto Guidoni
Yuri Lonchakov

E Apr 19-May 1/01 STS-100

Steven Lindsey (C)
Charles Hobaugh
Michael Gernhardt
James Reilly
Janet Kavandi

A Jul 12-24/01 STS-104

Scott Horowitz (C)
Frederick Sturckow
Daniel Barry
Patrick Forrester
Frank Culbertson
Vladimir Dezhurov
Mikhail Turin
James Voss
Susan Helms
Yuri Usachev

D Aug 10-22/01 STS-105

Dominic Gorie (C)
Mark Kelly
Linda Godwin
Daniel Tani
Yuri Onufrienko
Daniel Bursch
Carl Walz
Frank Culbertson
Vladimir Dezhurov
Mikhail Turin

E Dec 5-17/01 STS-108

Scott Altman (C)
Duane Carey
John Grunsfeld
Nancy Currie
James Newman
Richard Linnehan
Michael Massimino

Co Mar 1-12/02 STS-109

Michael Bloomfield (C)
Stephen Frick
Jerry Ross
Steven Smith
Ellen Ochoa
Lee Morin
Rex Walheim

A Apr 8-19/02 STS-110

Kenneth Cockrell (C)
Paul Lockhart
Franklin Chang-Diaz
Philippe Perrin

E Jun 5-19/02 STS-111

Jeffrey Ashby (C)
Pamela Melroy
David Wolf
Piers Sellers
Sandra Magnus
Fyodor Yurchikhin

A Oct 7-18/02 STS-112

James Wetherbee (C)
Paul Lockhart
Michael Lopez-Alegria
John B. Herrington

E Nov 23-Dec 7/02 STS-113

Rick Husband (C)
William McCool
Michael Anderson
Kalpana Chawla
David Brown
Laurel Clark
Ilan Ramon

Co Jan 16-Feb 1/03 STS-107

Eileen Collins (C)
James Kelly
Charles Camarda
Wendy Lawrence
Soichi Noguchi
Stephen Robinson
Andrew Thomas

D Jul 26-Aug 9/05 STS-114

Brent Jett (C)
Christopher Ferguson
Heide Stefanyshyn-Piper
Joseph Tanner
Daniel Burbank
Steven MacLean,

A Sep 9-21/06 STS-115

Steven Lindsey (C)
Mark Kelly
Stephanie Wilson
Michael Fossum
Piers Sellers
Thomas Reiter
Lisa Nowak

D Jul 4-17/06 STS-121

Mark Polansky (C)
William Oefelein
Joan Higginbotham
Robert Curbeam
Nicholas Patrick
Sunita L. Williams
Christer Fuglesang

D Dec 7-?/06 STS-116

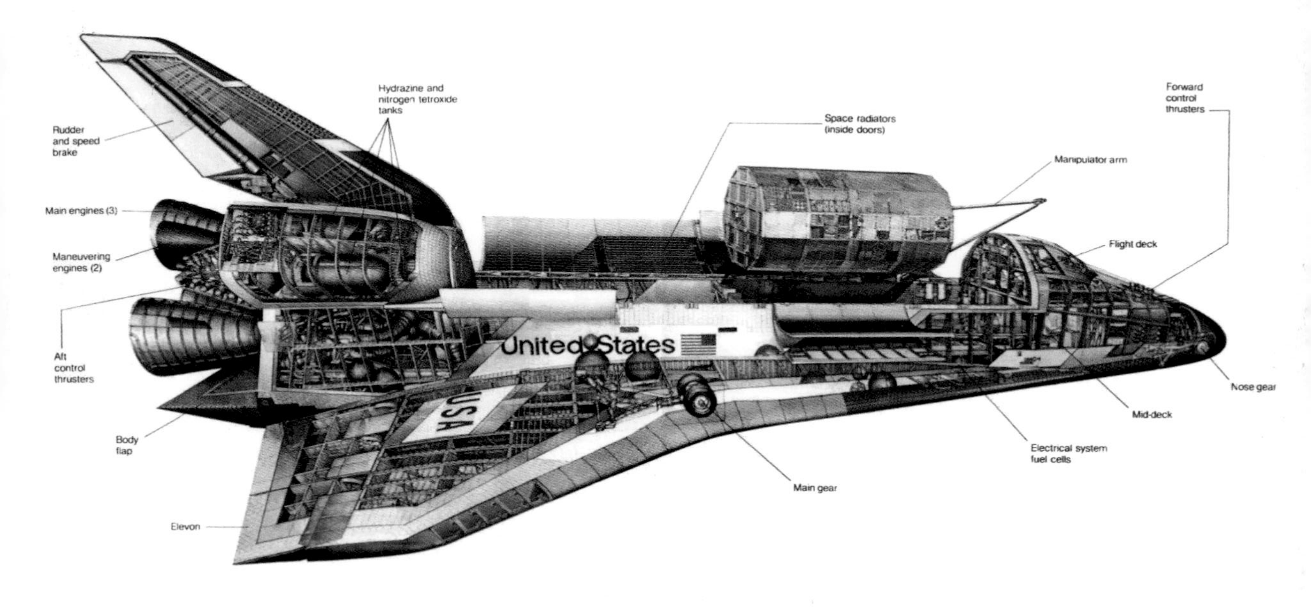
Rudder
and speed
brake
Hydrazine and
nitrogen tetroxide
tanks
Main engines (3)
Maneuvering
engines (2)
Aft
control
thrusters
Body
flap
Elevon
Space radiators
(inside doors)
Manipulator arm
Forward
control
thrusters
Flight deck
Nose gear
Mid-deck
Electrical system
fuel cells
Main gear
United States
USA

STS-67 rolls out of the Vehicle Assembly Building

STS-70 arrives at the Launch Site

The first space shuttle launch April 12 1981 note the white paint on the external tank

STS-27 Atlantis launch
December 2nd 1988

Solid Rocket
Separation

Startling image of
STS-121 taken
from the Solid
Rocket Booster
after separation

The ESA designed Spacelab is seen in the cargo bay

Challenger is photographed by the SPAS mini-satellite. Note two satellite pods at the rear of the cargo bay.

Perhaps the shuttle's greatest moment, the repair and re-deployment of the Hubble Space Telescope (above)

Satellite deployment and repairs were standard operations during the first few years of the Shuttle program. Here we see the crew of STS-49 repairing Intelsat.

STS-86 arrives safely back on the landing strip at Kennedy Space Centre October 6th 1997

Endeavour is processed by the team at Edward's Air Force Base while Columbia leaves for Florida. October 12th 1994